ARSON!

ARSON!

by **John Barracato**
with **Peter Michelmore**

W · W · *Norton & Company* · *Inc* · *New York*

Copyright © 1976 by John Barracato and Peter Michelmore
First Edition
Library of Congress Cataloging in Publication Data
Barracato, John.
    Arson!

    1. Barracato, John.   2. Arson investigation—
Personal narratives.   3. Arson—New York (City)
I. Michelmore, Peter, joint author.   II. Title.
HV8079.A7B28   1976   364.1'64 [B]   76–7383
ISBN 0 393 08744 1
All Rights Reserved
Published simultaneously in Canada
by George J. McLeod Limited, Toronto
This book was designed by Paula Wiener.
Typefaces used are Caledonia and Caslon Bold.
Manufacturing was done by Vail-Ballou Press, Inc.
Printed in the United States of America
1 2 3 4 5 6 7 8 9 0

This book is for Flo and Nan.

Contents

Some names and places have been changed in this book to protect identities. The cases are real.

ARSON!

Chapter 1

Cowboy
Charlie Brewer

THEY had their beginning, my street years, at a dinner party on Staten Island in the late winter of 1967. During pre-dinner drinks a barrel-chested man with a polka dot bowtie came over and introduced himself as Jerry Slattery.

"They tell me you're a fireman," he said.

"Yeah. Ladder Company 81. Right here on the island."

"Know it well," he said. "I'm with the bureau of fire investigation, a marshal, or as you guys in the firehouses would put it, 'one of the phonies.'"

"Not me, brother." The man had half a load on and I didn't want a hassle. What he said was true enough. Marshals were arson investigators who sometimes seemed overtaken with self-importance because they carried guns, drove around in unmarked cars and had full police powers.

"It doesn't matter anyway," Slattery went on. "I'm quitting. Goddam department expects you to work all hours of the day and night for a few dollars more than you guys get. And there's never any appreciation shown for what you do. Look, take my word for it, we have some

1

marshals, Charlie Brewer, for instance, who could match any detective in this city. Charlie's arrested hundreds of arsonists. He's the sort of guy who can ignore the lousy conditions we work under and just get on with the job, beholden to nobody."

"I've seen his name in the papers," I said. "In fact, it had crossed my mind once or twice that I would like to try out for marshal myself. I'm thirty-four, gotta make my move some time."

Slattery took a long pull on his highball. "You could probably get my slot if you wanted it. What about it? Want me to set up an interview?"

I looked at my wife, Flo. She had been seriously ill and was still on a strict regimen of rest and medication. How could I think about a job that would take me away fom her and out of touch for such long, irregular hours? She read my mind and gave me an emphatic nod.

"Thanks, Jerry." I said.

"Any time."

Within the week I was nervously describing my background to Chief Fire Marshal Vincent Canty in the old fifth-floor offices of the investigation division in the Municipal Building in downtown Manhattan.

Canty, a thin, red-faced man in his fifties, seemed unimpressed until I told him that I would be honored to be a fire marshal. Nobody could have laid it on quite that thick to him before, though, in truth, I was sincere. He gave me an application form. To qualify I needed some college courses, which, fortunately, I had taken in the Army, along with finishing my high school diploma. A second language was also required. I wrote "Italian." It was a lie but who would doubt a guy with Barracato for a name.

A few days later I was instructed to report to the Fire Department Training School, arson investigation course, on Welfare Island. For thirteen weeks, in a small class of twelve men, I studied fire examination, basic law, courtroom demeanor, report writing, evidence taking and a variety of other practical subjects that so intrigued me that I got a perfect score on the final test and graduated to a grade B provisional fire marshal. That was another chafing factor about the division, then called a bureau, in those days. All seventy-three marshals, even the A grades, were provisional. Two years later, during the period of reorganization when the Fire Department moved to Church Street, a department-wide examination was held to get a roster of permanent marshals, no grade As and Bs. Half of the provisionals flunked and were replaced by men from outside the division.

I felt six inches taller every time I rode the elevator in the Municipal Building and got off in the brown marble corridor that led to the world of the fire marshals. Most of them were ex-firemen in their thirties and forties, prowling the city in pairs, working their butts off. Official figures listed around a thousand suspicious fires each year in the five boroughs, but the true number was five times that and going higher. Marshals were required to investigate every suspicious fire and every major fire. If arson was established—and I could see then that many of these guys could pick a deliberately set fire as easily as a doctor could read a bone break in an x-ray film—it was their job to hunt down the arsonist. With their coming and going, and the telephones jangling all the time with fresh cases, the main division office had the frenzy of a giant newsroom. Down the corridor a storage room had been set up with old cots and lockers so that the men could catch a nap between jobs. It was bruited about that the division had the highest per capita rate of divorces and heart attacks in the city administration, and I suspected that Canty, who never missed morning Mass, prayed for the souls of his men. He realized that you could not push men as hard as the marshals were being pushed without them taking out some release time for mischief.

It was the system of the day to pair off the rookie marshals with senior A men, assign each team to a borough, and mix day and night tours within the same week. When the work sheet was posted for my first week I ran my finger down the list of men under Brooklyn, which I had requested. My heart jumped. I had drawn Charlie Brewer.

Brewer was already on the floor when I read the pairing and although I had met and spoken with him before, I approached him at his corner desk with some reverence. He was a lean, angular man with a hard, flat forehead, thin brown hair and restless blue eyes, a good twelve years older than I was, and the acknowledged ace of the bureau.

"You're lucky, kid," he said, grinning up at me. "I know this town like the inside of my pants pocket. It's some son of a bitching place, I can tell you. You'll be going into areas where they'll eat your heart out while it's still pumping if you let them."

As those early days and nights progressed, with Brewer taking me in tow while he investigated tenement fires and interrogated suspects in central Brooklyn, I had the bleak realization that I might never make it as a tough street detective. Brewer was the archetypal third-degree cop of the movies, threatening, slapping his suspects around, calling them animals, anything for a confession and an arrest. In the car, and when I took him home to meet Flo, he was full of fun and hilarious stories. In the precinct houses, where we did our booking and questioning, he was a cowboy.

One night we hauled in a Puerto Rican kid, about eighteen. Brewer was convinced that he had set a basement fire that destroyed two floors of the apartment house above, but even when Brewer batted him about the head the kid refused to change his story: he had been in the apartment house to visit friends and knew nothing of the fire until he had seen smoke and run for his life.

"Okay, Juan," Brewer said finally. "We have a way to get to the truth. Will you submit to a lie detector test?"

"Here?" said the kid.

"Sure," said Brewer. "I'll show you." He faced the kid to the wall and went off into a secretary's alcove adjoining the precinct squadroom. He hooked a wire coat hanger off the peg and bent it so that the hanging wire was a triangle shape. The hanger was jammed into the roller of a typewriter, rail hook down, and covered with a drape Brewer pulled off the window. Only the peak of wire showed when he wheeled his contraption back into the squadroom.

The kid glimpsed the wire and turned back to the wall. He was not about to let on that he did not know a lie detector when he saw one.

"You understand that if this wire glows red you're lying," said Brewer. "If it remains unchanged you're telling the truth. First question: what's your name."

"Juan Gonzalez."

"Look at that, John," Brewer said to me. "No change. He's telling the truth. All right, Juan, now you set that fire tonight, didn't you?"

"No," said the kid.

Brewer shouted with excitement. "John, how about that? The wire's glowing red as a stripper's nipple. We've got you, Juan. Let's have it all."

Incredibly, the kid confessed that he and two friends had gone into the basement of the building to sniff glue. They got high and set fire to a pile of empty cardboard cartons so they could pull the alarm and get their kicks from watching the fire crews perform.

Brewer chuckled all the way back to Manhattan. I wondered if I could ever dream up a stunt like that, and, if I did, would I use it or reject it as putting a suspect under duress. In any event it taught me the value of bluff—a subject that had not been covered in school and one I would use, though less outrageously, in the years ahead.

Everything I did with Brewer was in the nature of a postgraduate course. In time I would ask Canty to separate us because I was absorbing Brewer's reputation for bully-boy tactics, but in the summer of 1967 he dazzled me.

Arson was the worst crime, said Brewer, because it was

savagely indiscriminate. A landlord burning down an apartment house for insurance, or a jilted lover torching his girlfriend's house, could murder many innocent people—firemen among them—in the prosecution of their greed or vengeance. In the pursuit of arsonists, therefore, it was permissible to overlook the rules and improvise as one went along.

We were called in on a case involving the burning of an electrical appliance warehouse in Jackson Heights. The building was in disrepair and half the wrecked appliances seemed to be lacking essential parts, but the warehouse and contents were heavily insured and the owner stood to make a fortune. Another team had a professional torch in custody for the job. This man claimed he had been hired and paid by an Armenian-looking guy outside a grocery store in the Greenpoint district of Brooklyn. He did not know the man's name or address, but he gave a good description, and he agreed to identify him from a showup of pictures in exchange for a deal by which he would get a lesser sentence for his crime. Brewer knew the store as a numbers drop for an outfit controlled by one of the Mafia families. By this circuitous route we found ourselves climbing the stairs of a four-story brownstone opposite the store one morning with binoculars and a 400 m.m. lens camera. We intended to park on the rooftop and take pictures of everyone answering the torch's description in the vicinity of the store. If we found the right man he would be used in turn to put the finger on the owner, who was surely the main culprit in the arson.

"Jesus," Brewer complained as we climbed up. "We'll be here for a week. Half the people in this city look Armenian to me."

On the fourth-floor landing, in the dark shadows by the scuttle door leading to the roof, two shapes suddenly stiffened at the sound of our steps.

"Hold it! Don't move!" Brewer had his gun out and was by their side in an instant, frisking for weapons. I came closer to see a young man and woman, light-skinned Hispanics, he holding a hypodermic needle and she with her jeans down over her thighs.

"You got us," yelped the man. "This here's my wife. She's sick."

The woman was trembling, her face wet with perspiration. She made no attempt to pull up her pants. I assumed she was well into the agony of heroin withdrawal. The man had a spoon of the white powder and a candle stub on a ledge by the scuttle. Obviously he had been preparing to shoot her up in the hip.

"You animals are always at it," Brewer snarled. "You're under arrest. We're going to the precinct."

"Hey, keep it down, Charlie," I said. I realized we were required to arrest the couple, but I was surprised Brewer would make such a noise about it, jeopardizing our hopefully secret surveillance.

The man cowered and went timidly as Brewer pulled him and his wife up the short flight to the deserted roof. Brewer established that they lived in the neighborhood, a few doors down, and frequently visited the grocery store across the street.

"Fine," said Brewer pleasantly. "Now if you will cooperate in a little caper of ours we might just overlook what we saw downstairs."

"You be's the man. Whatever you say." The husband was still holding the needle and he had brought the other fixings without spilling a grain. "Can I shoot up my wife first? She's suffering, man. Ain't gonna do no harm to make her peaceful."

The woman lay back against the bulkhead. Every few seconds her body shuddered in violent spasm.

Brewer nodded. "Make it quick."

The husband lit the candle, moving the flame under the spoon. He sucked the heroin into the needle and injected under his wife's buttocks, playing with the plunger until blood kicked back into the cylinder.

"Drive it home, for crissakes," said Brewer.

"No, easy does it." The man was concentrating on his sickening task. "She's on a three-bag-a-day kick. There's only a bag and a half here. Gotta saturate."

It was a full minute before the needle was withdrawn and the woman slid down to the asphalt roof, her eyes half closed, a calm coming over her face and body.

We left her there and crossed to the parapet overlooking the street below. Brewer gave the husband the binoculars and instructed him to identify as many men as he could as they entered and left the double-fronted grocery market opposite. We snapped long-range pictures of Dr. Rasheed, the block dentist, a swarthy numbers runner named Gippo, a loan shark operator everybody called Mr. Vermicelli to his face and The Vermin privately, and a dozen others of a dark Eastern Mediterranean appearance. Towards midafternoon a green Cadillac parked along the street. A middle-aged man with strands of black hair plastered across the bald dome of his head got out and walked towards the store. Our spotter said his name was Haddad, the figures man who came every day to do the accounting of the daily numbers take. We finished with Haddad, leaving the Hispanic man and his wife on the roof. She had joined us hours before and I was startled to see how pretty she was—a doe-eyed girl with creamy skin, no needle track marks anywhere in sight.

At the showup of pictures—a gallery of men of similar appearance, as required by law—the torch identified Haddad as the man who hired him. We were able to tell the marshals handling the central investigation that Haddad was an accountant for the mob and probably did the books for the warehouse, now suspected as a drop for hijacked merchandise. The tip proved out; Haddad was indicted for conspiracy to commit arson. He flatly rejected a deal to go turnabout on his employers and did not appeal his sentence of one to five years.

Around August that year Brewer was called in for a long conference with Canty and when he came out to me at the marshals' desks his blue eyes were sparkling and he was rubbing his hands together as if he had been promoted to Fire Commissioner.

"What do you say, kid? Want to get in on a plant?"

"A plant?"

"Yeah, stake out a bad area, wear old clothes, real Hollywood stuff."

"Where?"

"Brownsville." Brewer spoke the word with relish. It was the toughest ghetto in the city, a rash of fires every night. We had been there often on investigations, but it had been quick in and quick out. I barely knew the area.

"It's a volunteer job," Brewer continued. "We split and work alternate weeks with a detective there. They say from four P.M. to two A.M. every night. That doesn't mean a damn thing on a plant. You get a guy in your sights and you stay with it. Canty says you're too green. I told him it'll be good for you to get your ass out on the street."

"Count me in," I said.

Chapter 2

Christopher Street Torch

BROWNSVILLE, where I received my true baptism in the rich adventure of the streets of New York, is a large neighborhood in east-central Brooklyn, bordered by Bedford-Stuyvesant on the north and Flatbush on the west. It was a mildly prosperous, predominately Jewish district up until the late 'forties. The rot set in then and by the summer of '67 it was generally recognized as a giant seething slum. Newspaper feature writers called it a darkly sinister cauldron of poverty, hate, and fear. A stranger who emerged from the IRT subway in Brownsville in the daytime was given a fifty-fifty chance of walking a block without getting mugged. At night he was given no chance at all.

On my first day, however, scouting the area by car on the way to the 73 Precinct, I did not perceive it as totally menacing. I sensed then what the years would teach repeatedly: no district in New York is without redemption. There are asphalt jungles, but there are many more good people within them than predators; perhaps in equal proportion, though on a different scale of value and violence, to the wealthier neighborhoods.

I found some streets in Brownsville shaded by big

plane trees, and here and there a fine apartment house, geraniums on the stoop, and friendly corner groceries. Pitkin Avenue, the main shopping street, had bright stores with new merchandise in the windows and good-looking women bustling in and out. But there were also many pockets of filthy tenements with ragged curtains showing through cracked windowpanes, and sullen men and women lounging about the sidewalks. There were burned out buildings by the dozen, rutted pavements, broken-down cars, and yards strewn with uncollected garbage and stinking with dog mess. And these areas of decay were spreading. Long-time residents who had some pride in Brownsville were moving out before a wave of immigrants from the south and from Puerto Rico. The newcomers jammed into the most dilapidated of the old brick, four-story tenements, paying high rents and getting no services in return. Every penny they made at menial jobs, or scored from welfare, went for food and shelter. They felt trapped there, totally alien from the leafy suburban world of the television screen. Nobody was giving a damn about them. Out of this plain human uncaring came the despair, the lawlessness, the compulsion of many to destroy.

Arson had become epidemic in Brownsville. Bitter young bucks found in arson a dramatic demonstration of their rage, but while they had razed numerous buildings and attracted the major publicity, they were not the only culprits by a long shot.

Landlords who saw a bleak future for their investments in Brownsville were hiring torches to burn their buildings for insurance. Tenants themselves were setting fires in their buildings to get relocation priority to the new city apartment developments, which were unavailable if you already had a place to live. Street gangs threw Molotov cocktails into hallways to even scores with other gangs. Thieves set fires to facilitate and cover their larceny. And like moths to the flame came the pathological fire-setters, the pyromaniacs. Their chances of getting caught while pursuing their warped pleasures were about zero.

My partner for the Brownsville arson plant was Larry
Goldman, a squat, pear-shaped, middle-aged fellow from
the 73 Precinct who had never made it above third-grade
detective, or senior third-grade detective, as he preferred
it. The low grade had nothing to do with ability. Goldman's misfortune was that he lacked a rabbi, someone to
sponsor promotion. During World War II he had been an
arson sabotage investigator for the government. His nos-
talgia for the work got him detailed, happily, to the patrols
with gung-ho fire marshals.

I took to Goldman right off. He knew the streets, and he
was unsparing in his advice about how I could escape getting my head beaten in. In years to come the things he
told me doubtless saved my life many times over. Nothing
he said then, however, could bring my wariness and exu-
berance into the right balance.

By staying up half the night on rooftops and sitting by
the hour on stake-out in one of the Fire Department's patrol cars, we made several arrests in the early weeks of
our tour. Some of the men we hauled in were obviously
hired for the torch jobs, though getting them to name the
bankroller was extremely difficult. They expected to beat
the rap, or get off lightly, and did not want to spoil it for
future contracts. One technique was to scare tenants out
of the building with small rubbish fires, for which the
torches would play good citizen and go pull the fire alarm
box. When the place was empty, the fire rigs gone,
whammo, a couple of jerry-cans of gas splashed about the
front hallway, a match, and the place would be an inferno.

It was a dangerous game. One arsonist literally blew
himself out the front door of a house on Walsh Street
when he tossed his match too soon.

When we arrived at the scene we could see by the way
the door was knocked back on its hinges that there had
been an explosion in the hallway. Whoever set the fire
would not have escaped the blast. We went straight
around to the Kings County Hospital casualty ward. A
Hispanic kid about nineteen years old was face down on a

cot having the singed remnants of his shirt and dungarees scissored away by a young Pakistani intern. The doctor shrugged when we showed our badges and went back to cutting, pausing frequently to remove fragments of cloth that had been welded to the black, burned flesh on the kid's back. We assumed he had been given injections to kill the pain, but he was still whimpering. The kid told us in short, gasped sentences that he had been burned when he was lighting the gas oven in his apartment. Suddenly, he said, it blew. We gently informed him that his back was burned, not his front, and we bluffed that we knew positively that he had set the fire on Walsh. From our observations there we even described how he had done it. The kid confessed then that he and two pals, who got out before he threw the match, had been paid one hundred dollars apiece by a complete stranger to torch the place.

"Suckers," said Goldman. "The going rate is a grand."

"That cheap bastard landlord!" The kid screamed the words in pain and anger.

Later, we had the district attorney's office guarantee the kid immunity in exchange for evidence against the landlord, who turned out to be a rich citizen of Westchester County. He was indicted for conspiracy to commit arson and reckless endangerment. God knows whether he ever served time; probably only a few months. It was our job to catch the arsonists and stick them with unbreakable evidence. Rarely did we follow a case all the way down the line. Ninety percent of arrested felons got off lightly by copping a plea, admitting to a lesser crime, and if you dwelt on this too much you could get ulcers.

We nabbed a few hired arsonists in the act of pouring gasoline down the scuttles on the tenement roofs, and a few more at the foot of fire escapes after they had done the deed and the building was ablaze. Street gangs used the roofs of Brownsville as readily as they did the streets, going up one fire escape and crossing the asphalt and parapets to a tenement five doors down. One trick was for an advance man to set a small fire in the back of the front hall then run up the stairs hollering the alarm. As the tenants bailed out of the building in their nightclothes and the overworked men of Ladder 120 and Ladder 103 came clanging up the streets, the roof squad ransacked the deserted apartments of television sets, stereos, cameras, anything they could carry. Goldman and I wised up to this maneuver and caught them on the fire escapes.

"Jesus, man, you crazy?" they'd say, puffing and perspiring under the weight of a 21-inch portable. "I live on the second floor. I heard the alarm and grabbed my stuff and ran."

Their stories usually came apart at the precinct house and when we could not prove the arson we had them booked for larceny. Junkies were involved in many fires, either as cheap torches-for-hire or as larceny-minded arsonists. Dozens of them operated on Herkimer Street, which was fast being reduced to rubble. The remaining residents were arming to defend their buildings from intruders.

Late one night we were called to a tenement there that had sustained three fires in the previous few hours. We jabbed at the buttons on the apartment mailboxes in the entrance alcove until somebody activated the front door release. At the sound of the buzzer we pushed on into the long, narrow hall. A low-watt bulb under a cheap pink shade cast sufficient light for us to see two apartment doors down the right side, the stairway to the left. We were making for the stairs when the door to the back apartment swung open and a heavy-set black woman, about fifty years old, appeared in the hall. She carried a double-barreled shotgun at her hip. It was leveled right at us.

"I'm gonna kill you before you kill everybody else."

Goldman hissed at me. "Don't move."

Don't move. I was petrified.

"Lady, lady, don't shoot. We're cops." Goldman could not keep the tremor out of his voice.

The woman raised the gun barrel an inch. I felt myself passing wind with fear. "Larry, let's run for the stairs."

"Stay put," he snapped. "Don't open your mouth. I'll talk her out of it."

Goldman pleaded and cajoled, but the woman did not move the gun.

I ignored his command and spoke up myself. "Look, lady, I'm a fire marshal, from the Fire Department. My job is to get the people who have been setting these fires and making you folks suffer. Here's my badge, lady. Fire Department."

Whatever she thought of cops, or of plump little Jewish guys claiming they were cops, I figured she must have respect for the Fire Department. If it were not for firemen she would not even have a roof over her head.

"Come closer," said the woman. "Show me the badge."

I walked down the hall with the gold shield extended in my fingers. She watched it to within a few feet of her, then lowered the gun with a heavy sigh.

"I'm sorry, man, awful sorry. You don't know what it's like living in a building like this. Look at my apartment. Water's dripping down from the ceiling over my furniture."

We did go into the apartment. It was neat, clean, a French antique style sideboard, big easy chairs covered with a floral print, family pictures on the wall. Three black water stains marked the white ceiling in the living room, and a trickle of dirty water dropped down from each onto a bright orange carpet.

The woman was still holding the shotgun as she showed us to the door. Goldman gently took it out of her hands, shucked out the two shells, and handed it back.

"We're sorry, too," he said.

For nights after that we prowled the derelict buildings in the vicinity, seeking out the junkies in the basement rooms they used as shooting galleries.

"We don't mind it if you shoot it up," I told them, talking tough. "But you knock off the fires or we'll have your balls busted."

Impressed at our street sense in knowing where to find them, and our arrogance in stalking Brownsville at 2 A.M., the junkies took the threat seriously. The last thing they wanted was to be locked away from their stuff. The fires along Herkimer Street diminished and finally stopped altogether.

My Charlie Brewer act did not impress Goldman. He had me pegged, correctly, as a boy-man embarked on a keen adventure.

One warm Monday in June, when I had rejoined him after a week off, we went to check out the site of a weekend fire on Christopher Street. It was the sixth fire in that block in the space of two weeks and we agreed with firemen in the local houses that they were caused by the same person. They called him "the Christopher Street torch," and wondered why two guys who had nothing else to do all day but catch arsonists were unable to bring him in.

Goldman pulled the Plymouth over to the curb by an abandoned four-story apartment house that had been tinned up by the city and marked for urban renewal. Metal had been torn off the lower windows by the ladder company men who had knocked down the weekend blaze, but otherwise it stood there intact as a blind monument to better days. Fancy cement work over the front archway indicated that the builder had really done a craftsman's job when they put up the place at the turn of the century. I could imagine that in those early times quite a few families had made a fine home of it.

I got out of the passenger seat and looked up at the higher windows; each had its own cement scroll capping.

"Don't ever do that, John." Goldman had come around the car and was wagging a finger at me.

"Do what? I wasn't doing anything."

"You were looking up. Don't ever stand looking up. If somebody throws a pot of lye down you'll catch it full in the face, maybe get blinded for life."

"Larry, the building's empty. Who's gonna throw lye at me?"

"Get the habit, John. If you're going to survive the streets you have to watch your head."

Goldman went on up the stairs. The sheet metal seal on the doorway had been wrenched aside and the door inside was tilted on its hinges. He pushed it open with his foot, peering into the hallway before he went ahead.

"Here it is again," he said. He was standing under the staircase poking at a burnt pile of debris on the floor with the toe of his shoe. "Same place. Same m.o."

The underside of the staircase was a honeycomb of charcoal and some of the struts and part of the bannister above had been burnt through. The fire had reached the second landing, scorching the timbers there, but then someone passing by must have seen the smoke and pulled the alarm at the box on the corner. Firemen had extinguished the fire before it had properly taken hold in the natural flue of the stairwell and gutted the building. Identically set fires in other tenements along the block had had much more disastrous result than this one.

We spent the rest of the afternoon tooling around Brownsville talking to informants, that strange breed of shifty men who hang out on the streets ingratiating themselves to hoodlums and policemen alike. They'd inform on their mother for a flipped half-dollar and a chance to call a pusher by his first name. And it was the same with cops. Whatever the department thought, the street people felt that my fat partner was the smartest detective in town. Informants felt privileged when he asked their advice, and they gave it eagerly this day. The Christopher Street torch, they said, was almost certainly Manny Castro.

If anything, the news was disappointing. Manny was a tiny Puerto Rican fellow, no more than four feet ten inches tall, who wandered the streets in an alcoholic stupor. He did not speak to anybody, and nobody bothered to speak to him. One of life's discards was Manny. He slept in deserted houses, a pint of Thunderbird at his side, not worth rousting by the most desperate of junkies. His extreme isolation from society marked him as a potential or practicing pyromaniac, as Goldman and I had both realized long before. For him the act of making a fire could provide the sexual stimulation unavailable any-

where else. Although knowledge of the connection between pathological fire-setting and sex was quite primi-
tive at the time, there were recorded cases of men and
women achieving orgasm over the licking flames of fires
they had created themselves.

Short of tailing Manny twenty-four hours a day, there
was little we could do but roll in on the Christopher
Street fires as soon as we got a radio call from the Brooklyn dispatcher. In fifteen succeeding fires, however, we
failed to find the little guy anywhere in the vicinity. And
when we questioned him later he predictably denied
having any information that could help us.

"Please, please," he said, clasping his hands together
and cocking his head to one side. "I do nothing."

He looked at me with sick brown eyes that fixed on
mine for a split second, then flicked sideways, back to
mine, then sideways again. Never up and down, always
sideways. "I make no trouble for no one" he said.

The frustration was getting to Goldman meanwhile, and
my lessons in street survival were becoming more emphatic by the day.

Often he had explained that the Brownsville blocks on
the north side of Eastern Parkway, which you reach via a
slight incline, were the meanest in the neighborhood and
deserved special caution. When we turned that way one
day during the Manny Castro period he gave me his cus-
tomary "Watch your ass as you go up the hill," and I truly
responded with a firmer grip on the wheel.

Half-way up Howard Street a skinny beagle dog ran out
from between two cars and I clipped him with the right
front fender.

I hit the brakes and stopped within a few feet.

"What the hell are you doing?" Goldman was shouting
at me.

"The dog. Gotta see what I did."

"Forget the dog. Get the hell out of here. We'll get
killed."

I bore down on the gas pedal and it was blocks before I

was game to ease up. Goldman was calmer, his voice quiet.

"John, you just have to understand the tension on streets like Howard. You think we're safe in these casual clothes and driving around in this plain old car. You know, part of the scenery. Well, there isn't a baby on Howard that doesn't make us right away as the establishment. Some of these people are full of venom for our kind, and maybe they have their reasons. At the least excuse they'd like to kick in our asses. Hitting one of their goddam dogs is the best excuse in the world. It's enough to start a street riot. Okay, so we stop back there and a few bucks start to make trouble. You pull your gun and tell them to back off and before you know it there's a big mob pressing around. 'Fuck your gun,' they'll tell you. 'You've got five bullets. There's two hundred of us.' Supposing you have time to call in a ten-thirteen distress. Before you know it you'll have units coming in from all directions. There's a good chance then that people are going to get their heads broken. People are going to get shot."

The pear-shaped cop who could not get promoted to second grade looked down at his hands and linked his fingers. "That's not our job, John, shooting people, or getting them shot."

I appreciated the scolding and thanked him for it. But the part that set me thinking was about how much we stuck out on the street.

An idea germinated and I put it later to Goldman when he suggested another night stake-out in the car on Christopher, near one of the apartment houses that had not yet been torched. The metal nailed across the entrance had partially lifted at the bottom, leaving at least a four-foot opening for anyone to crawl through. Our pyromaniac would not be able to resist hitting the place some time.

"We're known too well, the car too," I said. "I got a buddy in a firehouse on Staten Island who makes wigs and mustaches, all the disguise paraphernalia, as a hobby. I'll get him to fix me up so that I look like a bum and then

I'll be able to hang around a lot easier."

Around noon the next day I made sure the target building had not been razed during the night, then nonchalantly strolled into the park opposite. It was a nice park, plenty of trees and swings and slides. Lately it had become a junkie haunt and that spoiled it for the mothers and kids. I took a seat on a blue-slatted park bench near a hole in the heavy link fence and put a half-smoked, hand-rolled cigarette in my mouth. I did not light it. A Bull Durham cigarette may look like a joint; it doesn't smell like one.

I had taken tremendous pains to look authentic and was pleased when Goldman said he would stay on watch in the car on the distant side of the park so that he could come to the rescue if I were attacked. A new drifter in town was a certain candidate for an exploratory mugging by the junkies.

A flowing mustache matched my black hair perfectly and a fake chin beard, reaching to the sideburns, advertised that I had not shaved in many days. I wore dark sunglasses in the manner of addicts and carried a portable radio, also typical of them. My teeshirt, khakis, and sneakers were torn and deeply stained. The sneakers bothered me somewhat. Every junkie we had questioned in arson cases wore size ten Converses as if they came out of the same hijacked truck. Mine were Converses, but size nine. Overall, I was satisfied that I had made the transition from a lean, hard-jawed Italian-American fire marshal to a street Puerto Rican. My skin color was not a problem, for the summer sun had darkened a complexion that was Mediterranean olive in the first place.

At one o'clock two Hispanics came bobbing along the path and sat down next to me. The way I was dressed, they looked like my brothers.

The man closest gave me a nudge on the arm and said something in Spanish. I grunted and nodded, not understanding a word. I brought the radio closer to my ear to show that I was busy listening to music. They would not shut up. First one and then the other would jabber at me.

Finally, I faced them directly, blew out my mustache, and snapped, "Fuck off."

"Sheez," they said, almost in chorus. "The man."

And with that they were off.

Luck was running with me that day. Less than an hour later, mincing along Christopher Street, came Manny Castro. He paused within a few steps of the building I was watching, then suddenly crossed to the curb and whipped out his penis. The gesture startled me. Introverts are rarely exhibitionists. Manny must be drunk. He peed himself dry of his Thunderbird wine in full public view, zipped up, and quick as a mouse spun around and dodged under the tin and into the abandoned tenement.

My heart started pounding like it always does when I get excited. I knew Manny was our man. Now my timing had to be exactly right. If I went in after him too soon I would blow the whole case. He had to be actually setting the fire. There are only two ways to prove arson—catch them in the act or show exclusive opportunity. In this case I could not tell a court Manny was the pyro unless I saw it with my own eyes. There was one chance in a thousand that Manny was innocent and wanted to use the building as a bathroom. Sure he had peed. Maybe he needed a crap too.

Ducking through the hole in the fence and crossing the street, I took a position at the side of the building from where Goldman could see me through a break in the trees. Four minutes, five minutes went by. I raised a hand in prearranged signal and dived in under the tin.

My sneakers were soundless on the chipped tile floor of the hall. I went forward towards the back of the staircase, gun in hand. Manny was crouched before a two-foot pile of old carton fragments and newspapers. Flames jumped from two corners of the pile and at the instant I saw him Manny was lighting the third corner with a long stick match.

"Police! Halt!"

Manny leaped at my shout and was poised to take off until he saw the gun.

I moved him against the wall, frisking him for weapons, and tugged him toward the door to escape the smoke. He was speaking rapidly in Spanish, thinking I was Puerto Rican.

We stopped at the entrance, our eyes burning from the thickening smoke. The fire had caught the staircase timber and spat and hissed as blue flames curled around the old varnish. I could not go under the tin first in case Manny pulled away and tried a back escape. If I let hin through he would run for his life.

I knelt on the floor and yelled through the opening.

"Larry! Are you out there?"

"Easy, pal, easy." Goldman's arm reached under the metal and I passed Manny through.

Once outside I sprinted up to the fire alarm box at the corner of Riverdale. Good old reliable Fire Department. The engines came screeching along within minutes, and soon the street was a mess of fire hose spaghetti. It was a hellish blaze to fight because the sealed windows kept the smoke inside. Of the team of ladder men who crashed through the entrance and up the stairs to knock out windows for ventilation, seven were overcome by smoke and had to be taken to Brookdale Hospital.

It is characteristic of pyromaniacs that after taking their pleasure out of the fire creation they lose interest and ignore the ensuing bedlam. Manny could not have cared less as firemen staggered red-faced and choking from his blaze.

"You make big mistake," he protested to Goldman. My partner had handcuffed him and was marching him away from the fire. As I followed behind, Manny kept twisting back and tossing his hair at me. "I go into house for shit. Please, mister, a shit. I see the fire and I'm trying to put it out when this man comes with a gun."

I started to worry when he kept to this story during the booking at the precinct. It sounded plausible. I had thought of it myself. Manny had a chance of making it stick in court. Later, on Christopher, I got some back-up from two witnesses. Young Puero Ricans, who saw Manny run into the building and me go in five minutes afterwards. They were not eyewitnesses to the actual fire-setting, however, and neither was Goldman.

Determined to lock up this case in every detail, I went next to Ladder Company 120 to get the battalion chief's report so that there would be no conflict on times. On the way out I paused to speak to the house watchman who mans the glass booth just inside and to the left of the big overhead door. He was a big, ginger-haired man with the Irish brogue that you hear so often in New York firehouses.

"I was thinkin' of takin' the course for fire marshal meself," he said with one eye winking at my dirty clothes and unshaven chin. "But if you lads are dressin' like that, sure I'm not thinkin' about it any more."

I was about to reply when I heard the chief's car start up on the ramp outside the firehouse.

"And where do you suppose a respectable fire chief is goin' at this hour of the evenin'?" I asked, mocking his blarney.

The ginger-haired man's face turned white.

"The chief," he said hoarsely, "is upstairs in his cot."

Somewhere between "upstairs" and "cot" I punched open the small hatch hinged into the overhang door. The chief's car, gleaming red in the streetlights, was accelerating and close to the curb line.

"Halt! Police!" I shouted for the second time that day.

The taillights blazed a brighter red as the brakes went on, faded again, then I saw them coming straight for me. The car had been thrown into fast reverse.

I was flatfooted with fear for a second. There was no way to get back inside the door, no escape. With the rear bumper two yards off I took a mighty leap and landed on the car's trunk.

My only thought was to find something to hold on to,

and I threw myself across the roof to seize the reflecting dome light. The driver, meanwhile, had braked again and was bucking the car across the sidewalk and bouncing down the curb. Good Christ, I thought, what am I doing here? If the car gathered speed and swung around a corner, I would be thrown off on my head. With controlled panic I hooked my right hand and wrist around the dome light and reached into the driver's side of the car with my left. The hand felt soft flesh, a man's throat. I grabbed it savagely and squeezed with my fingers until I thought they would snap.

The driver must have let go of the steering wheel with both hands to tear at the clutching fingers and jammed his foot on the brake in unthinking reaction, for the car slammed to a halt on the sidewalk opposite the firehouse. I came sprawling off the roof on the driver's side, but I was still anchored to the man inside by the grip on his neck. In one flurry of angry action I jerked open the door, hauled out the man, and spreadeagled him over the hood. I pinioned his hands behind his back with cuffs and leaned against the car with a tremendous sigh of relief.

"Sonofabitch," I said to the night sky over Brownsville. It had been a long day.

The would-be auto thief was a brawny young truck driver who had escaped that day from the psychiatric ward at Kings County Hospital. Once the police came and took him away, I lost interest in the whole episode and went back to making sure the collar did not come loose on poor little Manny Castro. However much I sympathized with the crummy life he had been dealt, he was a menace on the streets.

I worked in disguise in the weeks following and had my whiskers on the day I walked into a cold, shabby room in Brooklyn Criminal Court for the preliminary hearing to see if Manny's case would go to the grand jury for indictment. A smart young lawyer from Legal Aid, dressed to the nines in a three-piece tweed suit, soft cream shirt, and Kelly green necktie, suggested to the judge in appropriately euphemistic legal language that his client was being framed. The defendant Manuel Castro had never before seen Fire Marshal John Barracato, now present in court, who brought the charges of arson in the third degree (setting fire to an unoccupied building). In fact, I was not even the man who had mistakenly apprehended the defendant at the scene of the Christopher Street fire.

Goldman could testify otherwise, but at my turn on the stand I indulged in a piece of theater. Manny was the inspiration, throwing plaintive glances at the bench, shrunken in his chair like a small whipped dog.

With the court's permission, and with suitable flourish, I ripped off my mustache and beard, and confronted Manny as the true Barracato, the fire cop whose face had become know to every dude in Brownsville.

Manny came up in his chair and his mouth dropped open with a gasp heard around the room. The reaction sabotaged his lawyer's thin strategy. After much whispering at the table the defense rested and the case was duly remanded to the grand jury. We got the indictment all right, though Manny evaded trial by copping a plea for arson four (recklessly setting a fire) and spent only thirteen months in prison.

This I know because years later, in 1972, I was investigating a fire further down the same Christopher Street. The fire originated in the first-floor public hall of a two-story frame building and was caused by a Molotov cocktail. A charred wick and pieces of broken bottle were scattered about the floor. There was only one occupant of record in the place, a party on the second floor. In answer to my knocking, who should open the door but Manny Castro.

"Manny," I said, surprised. "What are you doing here?"

One look at his face and the man was down on his knees.

"Please, please, I don't even smoke any more."

"C'mon, Manny." I could not help laughing. "I realize this isn't your bag. You don't throw Molotovs."

We sat in his seedy apartment for the best part of an hour rapping about street gangs and their wanton Molotov habits. He told me about prison life, and I sounded off a bit about the current state of arson in the Borough of Brooklyn. Roaches came out for what seemed to be their winter games while we talked. They jumped from wall to table, from table to feet. I brushed several from my shoulder and squirmed at the suspicion that one had dodged down inside my collar.

I felt pleased about the encounter. Manny bore no grudges, and I had no antagonism toward him. I liked to think that we had a tacit understanding that the forces in my life had put me on one side of the law and the forces in his had put him, in his peculiar way, on the other. He had been a menace on the streets; it was my job to get him off.

The meeting reinforced an attitude that was germinating on the streets of Brownsville—that I would treat arsonists as I would like to be treated myself if I were on the other side. Many were animals, as Brewer would say, living day to day, without the stout heart to fight honorably against their social and economic trap. They were concerned only with getting their next fistful of money, a fix, a woman to rape. Others, however, were sick and their fires were more a cry for help. Either way it seemed too personally corrosive to go in with bare knuckles. Brutality was expected and put up as many barriers as it knocked down. Goldman's mild tactics were as effective as Brewer's harassment, I was to learn.

The greenhorn was at the beginning of perception that the so-called meanness of the streets was sustained and deepened by an emotional and physical chain reaction. Violence fed on violence, hate on hate, fire on fire. I did not want to be part of that lock-step. I vowed to defend my little store of nobility against the vilest extremes of the street—extremes to which I had had the slightest exposure up to the time I met Manny Castro.

Chapter 3

"Larry,
We Got a Roast"

AFTER a dinner of London broil and good Ruffino wine at Leo's Restaurant on Pennsylvania one hot Thursday night, Goldman headed our car into the East New York neighborhood for a change of scenery. We were talking over the stake-out possibilities for later when the radio crackled.

"Attention Brooklyn Box 1631. Fire reported in occupied dwelling. First floor. Corner Freeman and Dudley. Acknowledge all first due units."

The Brooklyn dispatcher's voice was unhurried, detached. I pictured him in his great barn of a control room over on Empire Boulevard, sipping a third cup of coffee to steel himself for a long shift of jangling telephones and flashing console lights. All fire alarms in the borough were recorded and dispatched from his communications center.

Answering calls came in rapidly.

"Ladder 120 responding."

"Engine 235 responding."

Goldman crushed his cigarette in the pull-out ashtray at the dashboard. "What do you say, John. We're only three blocks away."

"I'm with you," I replied. I unhooked the telephone-style headset from the radio and pressed the transmission button.

"Car 892 responding to Freeman and Dudley."

I grabbed the magnetic bubble light off the floor, plugged the heavy black flex into the cigarette lighter socket, and plopped the light on the roof of the car with one swoop of the arm. Goldman had flicked the siren switch and we were off to the fire.

Goldman roared up Freeman and stopped the car on the sidewalk at the Dudley corner, right outside a handsome dark brick apartment house with fresh white trim and ivy growing up the north wall. We were only seconds behind the ladder company. Firemen were still stretching their hoseline through the front door.

Inside, we saw smoke puffing out from underneath the door of the front left apartment. A stooped black man with grizzled gray hair was pointing at it and shaking his head furiously.

"I called the alarm," he said urgently. "I'm the super. Live in the back. You gotta be careful. Nasty woman in there, real nasty woman."

His words made the hair prickle on the back of my neck, for I had caught the unmistakable pungent odor of burning human flesh.

"Larry," I muttered. "Think we got a roast."

The superintendent's key released one lock on the door, but it held, bolted on the inside. A big ladder company guy motioned everyone back and crashed the door wide open with a single accurate blow from his axe.

A fireman with a portable, water-charged extinguisher led the way into a small hallway, through a kitchen, and into what was apparently a living room beyond.

"My God!" The cry from the advance firemen brought us rushing to his side. There in the living room, sprawled on the carpeted floor, was a man completely engulfed in flames. Firemen leaped to him in an instant, rolling him over and over on the floor until the flames were suffocated. The fireman with the extinguisher was concentrating on a stuffed armchair smoldering on one side of the room, and I noticed then, in front of the chair, an unlabeled one-gallon, silver metal can.

"Marshal!"

It was more a croak than a cry. It came from one of the fire crew who had crossed the room and opened the door to an adjoining bedroom.

I joined him there and stood speechless with shock.

Sitting on the edge of the bed in a filmy pink slip was a slender black woman in her early thirties. She was cradling a quart bottle of Old Grandad bourbon whisky against her breasts. She eyed us fiercely for a moment, then lifted the bottle to her lips, took a good long swig, and wiped her mouth with the back of her other hand. She put the bottle back against her bosom and starting shrieking.

"You got some goddam nerve breaking in my goddam door! I'll sue this fucking city for breaking down my door!"

The fireman found his voice first. "Who's that in the other room?" he asked evenly.

"Whadda hell do you care," said the woman. Her words were slurring. She was drunk.

A light went on in my brain, commanding me to get to work. I disregarded the woman completely. A rapid check of the apartment showed that all the windows were locked and that the front door was the only way in or out of the place. Since this had been bolted from the inside, we had a case of exclusive opportunity against the woman. The metal can on the living room floor gave off a strong smell of paint thinners, so we took this, along with the intoxicated woman, to the 73 Precinct. The man on the floor, more dead than alive, was removed by ambulance to Kings County Hospital.

We hustled the woman through the drab green doors of the venerable precinct on East New York Avenue, nodded to Lou, the duty desk sergeant, and half carried her up the metal stairs and along the passageway to the squad room. There was a detention cubicle off to one side, and we unceremoniously dumped her in there and raced back to our car. The man would surely expire; we hoped for a death-bed statement. About all we learned from the woman was that her name was Cora Sue Fricker.

At the hospital we parked in the "physicians only" lot and barreled right into the emergency ward. Our badges were pinned on our outside lapels and we were admitted without question. In New York City the hospitals cooperate with the law one hundred percent.

The victim, a young, thin-faced Negro, was naked on a table in a curtained-off alcove, his twitching body fed by tubes from three overhead bottles. Surgeons snipped away at the black-scorched flesh on his arms, chest and abdomen. I turned away, gagging. He was in no condition to answer questions.

A nurse motioned us outside the curtains and produced a fragment of his trousers, a charred belt, and the singed seat of his pants.

"It's all that's left," she said.

Goldman lifted the tattered remains of a wallet from the hip pocket. The driver's license said the man was Gerald T. Manning, of a Brownsville address. I found a corresponding listing in the telephone book and called the home. Mrs. Manning, the man's mother, answered and I told her where her son was and, briefly, what had happened. I spared her the detials.

Back at the precinct house we booked Cora Sue Fricker on suspicion of murder and arson, both in the first degree. She kept swearing that she was alone in the apartment and had no idea how a man came to be on fire in her living room.

"What about the paint thinners, Cora?" I asked her.

"What about it. When that goddam landlord painted my place he put more paint on the fucking carpet than he did on the walls. I got the paint thinners to clean up the mess."

"It won't work, Cora. You're booked and you're going to be fingerprinted and taken over to the 78 Precinct where there's a holding cell for females. Tomorrow morning we go to court."

"No way you bastards gonna fingerprint me. I didn't do nothing."

"Hey, Cora, you know better than that. No fingerprints, no bail."

After we took the prints I drove the route to the 78 while Goldman sat in the back seat. The handcuffs were in front of her so she was able to reach up and touch my shoulder.

"Gotta cigarette?"

I plucked a packet of Marlboro from my shirt pocket and handed it back.

"Jesus," she said. "You smoke these junky things. I only like Camels."

She clapped it in her mouth nonetheless while Goldman came up with a match.

"I'll light it, Cora. You're liable to set me on fire." Goldman chuckled at his joke.

I felt a surge of admiration for this wily cop. He was a genius at throwing people off guard.

Cora Sue Fricker came right in after him, leaning back in the seat, her voice casual.

"I don't know," she said. "New York is crazy. I did this once before in Oregon. They didn't make no big deal out of it."

We let it ride and drove the rest of the way in silence. A spontaneous admission, which this was, could be used in court.

It was around midnight when I got home to Staten Island that night. The bedroom lights were out, and I walked quietly upstairs like I always do to make sure my family was sleeping safe and sound—first Flo, then John and Dena. I went downstairs and put on the kettle for a cup of instant coffee. For two hours I sat in the living room, ramrod stiff on the edge of the couch, nursing the long-drained cup. The events of the night went around and around in my mind, never coming together to form a whole: the man on fire, the woman drinking from the bourbon bottle, the dim image of her slack face and tousled hair in the rear-view mirror, complaining as if she was being unfairly hassled for parking her car in front of a hydrant or something. There was no category in my experience for much monstrous savagery, and none either for its casual acceptance.

I finally went up to bed and fell into a troubled sleep that bordered on nightmare. A man in flames lay screaming and thrashing on the floor and I stood paralyzed before him, unable to help.

Goldman and I met at the hospital at ten the next morning. Manning had been removed to the burn unit, where his condition was listed as critical. The attending physician, a round-faced man with thick-lensed spectacles, said the patient was bleeding through his catheter, which, he said, meant kidney failure. Manning was not expected to live.

The doctor agreed to let us try interrogation, though he thought the exercise would be futile because Manning was heavily sedated. I went to the head of the cot and put my mouth close to Manning's ear, remembering as I did so the rules for an admissible death-bed statement. The victim must state his belief in God or a hereafter, must confirm that he knows he is going to die, must state the identity of his attacker.

"Gerald," I said quietly. "Can you hear me?"

The man made no response.

"Gerald." At this, his head turned towards my voice.

"Do you believe in God or a hereafter?"

His eyelids flickered for a moment, but did not open. His voice was weak and distant. "Yes."

"Gerald, you're very sick. The doctor relates to us that you are going to die. Do you know that you are going to die?"

Manning did not answer. His face, a muddy color, showed no reaction. I was standing by the bed, bending over him, as taut as wire. My hands, resting on the bars of the cot, became moist with sweat. I had to have more. My voice quickened.

"Gerald, Gerald. Do you know you are going to die? Do you know who did this to you?"

Again the flickering of his lids. "Yes," he said.

"Can you tell us what happened?"

His voice was barely audible. "Cora and me, we be's drinking all day. She says, 'Gotta have more whisky.' I got de whisky from de store, 'bout seven. I'm drinking whisky in de chair. Must 'a' passed out. Den I feel sumthin' cold and wet splashing on me and I see Cora standing dere . . ."

Manning's leg jumped. His tongue flicked over his lips. "Water. Cora's throwing water at me. She's reviving me. But I'se burning up, burning up."

Manning remembered nothing more of the hideous incident, but his statement, plus the exclusive opportunity, seemed sufficient to me to nail Cora.

We had her arraigned that day in Brooklyn Criminal Court and lodged in the detention cells there. As the bailiff was leading her from the courtroom she motioned me over from the front bench.

"Do me a favor, honey," she said. "Bring me back some Camels."

The request was in such mundane contrast to everything that had happened that I was angered by it. I strode from the court and down the broad stairs and out into Schermerhorn Street. Yet I paused outside the drug store on the corner and asked myself, why not? The woman was in custody, justice was taking its course. My personal feelings were irrelevant, or should be. I had been asked a small favor, and my normal reaction would be to go ahead and do it.

I went into the store, bought two packs of Camels, and left them for Cora with the female security guards at the main detention desk in back of the court building. At various times in the months following, during the hearings and leading to the jury trial itself, I must have given her a dozen packs of cigarettes—even, on a couple of occasions, a few dollars.

I kept the donations secret from Goldman and my fellow fire marshals. Cora was demonstrably a callous slut, with a yellow sheet of convictions a yard long. She was originally from Portland, where she had served time as a prostitute and also for assaulting her johns with knives, bottles, and, once, a pitcher of lye flush in the face. This must have been the case she referred to in the back of the car that first night. Apparently Cora went mad with the whisky. In jail in Brooklyn, away from the booze, she caused no trouble whatsoever.

The victim, meanwhile, lapsed into a coma from which he never recovered. We inspected the body, as required, in the Kings County Morgue. One of the pathologists slid out a huge steel drawer and there was Manning, naked on a slab. Most of the front of his body was burned away so that the bones of his rib-cage were clearly visible. Our flesh crawled at the sight and I remember Charlie Brewer, who had accompanied us, sucking air through his teeth.

"Holy Christ," he whispered. "This man really suffered."

Brewer was not a man to give way for long to morbidity. His gaze, and mine too, had lingered on the man's genitals, oddly untouched by the fire and now bunched enormously from his flanks.

"Hey, doc," Brewer said heartily, turning to the pathologist. "Any chance of getting a transplant? This guy's really heavily endowed."

The crude remark started us giggling like schoolboys and none of us dared examine our behavior for fear of bringing back our sickening dismay as we walked down the aisle of silver drawers and out into the crisp sunlight.

I was no match for Cora's lawyer at the criminal trial.

He stood there, a black man, tall and straight, tapping the stem of his reading glasses against his teeth, while methodically tearing my evidence to shreds.

The substance of his defense was that Cora's so-called spontaneous admission in the car was so vague as to be meaningless. The exclusive opportunity related equally to defendant and deceased.

"And what of this *death-bed statement* of yours, marshal?" He spoke in a mocking tone as if I had picked up the phrase from a James Cagney movie.

"I quote from the transcript: Question—'Do you know you are going to die? Do you know who did this to you?' Answer—'Yes.'

"What was the deceased affirming, marshall? That he was going to die or that he knew who did this to him? And what, precisely, was the 'this' referring to?"

During the summation, when the lawyer put it to the jurors that Gerald T. Manning, an alcoholic young man, may well have torched himself in a fit of despondency, I was tempted to jump up waving Cora's yellow sheet and read the catalogue of her vicious crimes against hapless johns who sought the pleasures of her body. The cause was lost, however, and Cora Sue Fricker walked out of the courtroom a free woman.

I was bitter about the outcome and mooched around division headquarters for days afterwards. Charlie Brewer snapped me out of it when we were cruising down Bushwick Avenue for a Brownsville patrol a week after Cora was acquitted. Our mentor, Larry Goldman, was on leave.

"None of the brass we got now gives a fuck whether you louse up a case or put fifty torches away for life," he growled. "Did you see anyone give Steve Stapleton a medal when he built an iron-tight case around that landlord in Rego who burned his hotel to the ground? The only one who got excited was Steve. He's his own man; he knows what he did. Did anyone tell you that I had turned down a three-quarters disability pension to stay on as marshal? No bastard understands why. I'll tell you.

Every time I bring in an arsonist I know that's one less guy out there trying to make this city shittier than it already is. I do it for myself. Fuck the department. Okay. You did a thorough job on the Cora case. The bitch should have been put away for a hundred years. But we got no time to fuck around debating the pitfalls of the American legal system. That big black dude showed reasonable doubt. So be it. Now let's get to the million-and-one arsonists in this goddam town."

I could feel his eyes boring into my right ear as I kept my face doggedly fixed on the road ahead.

"You gotta toughen up, John-boy, or they'll have your balls in a wringer—all of them, the hoods, the lawyers, the judges, the cops, the newspaper guys, everyone."

"Okay, okay." I wanted the lecture to finish in case he had somehow discovered that I had supplied Cora with cigarettes and he mentioned *that*. Fortunately, Brewer's interest had been taken by a car pulled over to the side of the avenue and a man walking away from it with a red gas can in his hand.

"You heard that one, didn't you?" he said. "This guy runs out of gas and takes off up the road to the service station. When he comes back there's a guy under his hood. 'Hey, pal, what are you doing?' he asks. 'Listen, man,' replies the guy under the hood. 'You can have anything you want. All I need is the battery and generator.' "

I laughed uproariously as I pulled off Bushwick, took a little piece of Eastern Parkway to Atlantic, then made a right on Howard.

"Watch your ass as you go up the hill," I yelled.

Brewer chuckled at the crack. "Jesus, that Larry," he said. "Some helluva cop."

Around eleven-thirty, the Brooklyn dispatcher instructed us to respond to a suspicious fire in a grocery store at 362 Madden, over in neighboring Canarsie.

"What's the reason for the suspicion?" Brewer asked on the radio.

"Molotov cocktail through the window," came the dry response.

The fire rigs had already departed when we arrived, leaving only a spoor of wet pavement. The grocery store was one of those narrow-fronted, stock-everything convenience shops which make the owners about eighty bucks a week. It was squeezed into the lower left side of a five-story apartment house, and cornered onto an alleyway. The area around the store was deserted except for an old station wagon parked down the side. Three boys, no more than sixteen years old, were sitting inside. We gave them a miss and went inside to talk to the shopkeeper, a neat little Puerto Rican man of about fifty-five.

"Such trouble," he said, more in the manner of a Jewish candy store man. "Such a night. Somebody threw a fire-bomb. Through the bathroom window. Why? Why here? I have no enemies. Complaints about the prices sometimes. But no enemies."

The biography of the fire was easily read. A Molotov cocktail had been tossed through the glass panel of the bathroom door which faced onto the alleyway at the back of the store. Gas-fueled flames had whooshed across the tiled floor, spreading up the walls and eating into the timbers of the ceiling and the apartment house floor above. If the fire trucks had not come then the whole building would have been ablaze. By the size of the place and the lights still shining in every window we estimated that fifteen families had been imperiled.

The shopkeeper said he had been upstairs at the time, watching the Channel 4 news. He understood that the boys outside had called in the alarm. Brewer stayed to take a statement from him while I went outside and stuck my head in the window of the station wagon.

"Hi, you guys, I'm John Barracato, a fire marshal. What's the scoop here? What happened? I'm told you guys pulled the alarm box. Real nice work."

The oldest boy, who said his name was Angel, said he

and his pals had spotted the fire from a friend's house, turned in the alarm, and run into the building to get the people out.

"Uh-huh," I said. "For my report now, where exactly were you when you first saw the fire?"

"Down the block," said Angel.

"Show me," I said.

We trouped off a good fifty yards to a tenement stoop.

"Here," said Angel. "We just finished playing pool in the basement, came up here for a cigarette, and saw thick black smoke coming from the direction of the store. We ran down the street and saw the flames. Joey here nearly got hit by a bus going across the street to the alarm box. Sam and I raced upstairs getting the tenants out. Man, it was some fire. We were coughing on the smoke."

Mention of a bus lodged in my mind but Angel had mostly lost me when he talked about thick black smoke. According to his time sequence the boys caught the fire right at inception. And a Molotov cocktail throws a bright flaming fire, almost no smoke.

We returned to the store, where I rapidly briefed Brewer, and then loaded the three teenagers into our Plymouth and took off for the local precinct. The desk sergeant glanced at our badges and waved us through to the squad room. Brewer telephoned the parents while I talked to the boys, one at a time, in a cramped squalid anteroom. Judging from my experience, the interior of every precinct house in New York City is painted two-tone green—mint and bottle. Nothing is more depressing.

"Joey," I said, picking the smallest of the three. "You three guys did a heroic thing tonight, saving those people. But why did you set the fire in the first place?"

"We don't make the fire," said Joey, his eyes popping.

"Sure you did, Joey. I know about these things. It's all I do, investigate fires. The city pays me to get the true story."

The city got its money's worth that night, for Joey, Sam, and Angel all came clean. The shopkeeper had suffered a burglary one night. Someone busted a hole through the thin wall between his store and the front entrance hall of the apartment house. He accused Angel, the cheekiest kid on the block, of being in on the heist. In resentment, Angel had gathered his two young, adoring buddies together, made a Molotov, and tossed it into the store bathroom. As they ran out of the alleyway they saw a bus passing by, and the driver staring directly at them and the flames shooting out behind them. Panicked, fearing they had been caught in the act, they dashed around to the front of the building and rousted the tenants. It was the bus driver, we later established, who had turned in an alarm.

"Maybe, just maybe, you'll make a good detective some day," said Brewer as we drove on back to Manhattan in the Brooklyn dawn.

I advanced a half step closer to that desired state six nights later when Brewer and I were diverted from an overnight watch in Brownsville to an apartment house fire in Forest Hills, Queens. A fireman had been critically burned. This, plus a feeling among the fire crews that there was something distinctly odd about the fire, brought us into the action.

The location was one of those idyllic sidestreets near the West Side Tennis Club, very English in architecture and atmosphere. Several fire rigs were still on the street when we rolled in about 4 A.M. and dozens of people were clustered about, some in nightclothes. The face of the dwelling, a three-story, six-family, red-brick place, was intact, but the interior was a smoking ruin. One of the engine company lieutenants, we learned, had been caught inside and was close to death in a nearby hospital.

Before we went into the building we moved among the firemen, putting together a picture of the state of the fire when they arrived.

"Damnedest thing," one of the engine company men told us. "We had a telephoned alarm but when we arrived and walked into the front hall we couldn't see or smell any smoke whatsoever. The lieutenant, Trevor Callaghan, went on up to the third floor by himself to see if there was anything there. We were about to call in an MFA * when there was an almighty roar and the fire exploded out of the basement and up the stairwell in a flash. A backdraft. Fortunately, the hoselines were ready and the ladder companies had arrived. We got the fire under control and the people evacuated. But Callaghan was in the path of the fire and he got it full blast."

One of the other firemen had a notation that the resident who called in the alarm was William Scott, who lived on the second floor.

"You question him," said Brewer. "He's bound to be here someplace. I'll talk to some of the other people."

We had not made a physical examination yet, and the fire could have been accidental, but we wanted more basic information.

Neighbors pointed out Scott as the tall, broad-shouldered man with crew-cut sandy hair smoking a cigarette and leaning his back on a parked Oldsmobile a few yards down the street. I went over and identified myself.

"When the smoke clears, I think you'd be well advised to check the wiring in the basement, marshal," said Scott, not shifting his position. I smelled the faint odor of whisky on his breath, but his voice was strong and precise. "They got a washing machine and dryer down there feeding off regular wiring. I told the jerk who owns the building to put in heavy duty, but he's the type who'd steal pencils from a blind man."

"Thanks, Mr. Scott. We'll check it out. When did you call in the alarm?"

"I guess about an hour ago. I'd just gotten in. Few beers with the boys, you know. What the hell. I live alone. Anyway I got to bed and I'm dozing off when I smell smoke. I put my head out in the hallway and there's smoke coming up from below. I telephoned the

* Malicious fire alarm.

emergency number, pulled some clothes on, and went down the back fire escape."

"Good. Thank you, Mr. Scott."

I left him by the Olds and went to find Brewer. "Charlie, talk to that big guy over there, Scott, and tell me what you think."

Brewer was back in five minutes. "The sonofabitch is lying."

We returned to Scott and asked him to step into our car to help us with a preliminary report.

"You're lying to us, sir," I said when he had settled in the back seat. "There is no way in hell you could have smelled smoke on the second floor at the time you said you did. We're not saying you set the fire, of course, but your story doesn't match the facts."

"I was half asleep, I told you that." Phlegm came into his throat and he cleared it away nervously. "I heard something crackling and maybe I just thought I saw smoke."

"Hey, hey," I said. "You're going in deeper and deeper."

Scott put the heels of his hands to his forehead and rocked his head back and forth. "Oh, Jesus, what a fuck up. I'm to blame, I guess, but it was an accident. The whole place burning down was an accident, I mean. I was drinking beer over at the Slipper on the Boulevard. It's only a few blocks away. I started to feel bloated so I switched to bourbon. They close the joint at two and I'm left with a double still in front of me. I carried it home in the glass, sipping it on the way, like we used to do in college. I was defensive end for Syracuse, for crissakes. It was the one drink too many because when I came into the hall and saw that goddam Gottlieb baby carriage I flipped. They live on my floor, Jewish couple with a baby that cries nonstop. Kid's nearly a year old and he's bawling night and day. Jews don't understand disciplining a kid. Jesus, what stupid people. Forest Hills is full of them. I see the carriage and for some insane reason I wheel it to the end of the hall and push it down the steps to the basement. Fucking thing is so expensive and wellsprung that it goes down the stairs and runs out on the floor like it's some goddam baby carriage ramp. I follow it down and kick the shit out of it. We stack old newspapers in the basement, for the Boy Scout pick-up, you know, and next thing I'm tearing sheets of paper from the piles and stuffing it under the carriage. I put a match to it and then I thought the hell with it and went upstairs. It's all cement, the basement, and I didn't think it would do any damage. But when I got to bed I kept thinking about it. What if it caught the workbench and the other junk in the basement? To be on the safe side I called the fire department and told them who I was and said that I thought there was a fire in the apartment house. I went back to bed until I heard the fire engine and . . . well, you know the rest."

Brewer, the man who had heard everything, stared at Scott with his mouth open. It was left to me to go into the formal speech: "You are under arrest, Mr. Scott. The charge is arson in the second degree. You have the right to remain silent . . ."

My partner let me finish, then came in with a voice like acid. "A fireman was burned over half his body tonight, Scott. A lad with a wife and three children whose job it is to clean up after fuckers like you. He's in hospital, in agony, Scott, and if he dies you're gonna be charged with murder and I will personally see that you fry in purgatory for the rest of your life."

Brewer was denied again, as he doubtless realized he would be. Trevor Callaghan did die and Scott was charged with murder. He raised the twenty-thousand-dollar bail out of his own pocket and was acquitted at the preliminary grand jury hearing. Brewer and I were never called to give evidence.

Chapter 4

Pitkin Avenue

THE ugliness I encountered in Brownsville during my first year as fire marshal was compensated by the thrill of the unexpected and also by a wild grain of humor that ran through a good part of the life there. It has always seemed gross to me in the telling, yet the low comedy does exist and it helps break the tensions and restore a perspective. Fire marshals spend nearly all their working hours in the lawdriest streets of the city. If they cannot enjoy some of it, they cannot endure.

One day shortly before Christmas, 1967, Larry Goldman turned the car into Pitkin Avenue, parked, and announced that it was time to meet some of the storekeepers who liked cops sufficiently to give them privileged customer discount. I would be buying gifts for the family, he said. Let's keep the money in Brownsville.

Goldman led the way along the sidewalk, his short legs pumping out from his rotund belly. "When I was a kid," he said, "people came from all over Brooklyn to buy at the stores here. Jewish families, and they were mostly Jewish hereabouts, used the avenue as the grand promenade. Ah, life was rich. It was the life of the *shtetl* in the old countries of Eastern Europe. Gone it is now, of course, though not entirely."

Goldman wheeled into an appliance store, the windows

43

jammed with cameras, radios, television sets. "Come," he said, "I want you to meet Eugene Rosenbloom."

Inside, white and Puerto Rican salesgirls were attending several customers. The boss seemed to be the Japanese leaning against the write-up shelf behind the main counter. He was smoking a curved briar pipe and wearing a deer-stalker hat, no less.

Goldman marched right over to him, extending his hand. Then he hailed me over and introduced me, without a flicker of expression, to Eugene Rosenbloom.

We were having a discussion about discount prices on cameras, me and this Oriental Eugene Rosenbloom in the deer-stalker, when Goldman interrupted: "Hey, Eugene, it's okay to show John some of your pictures."

With that, Rosenbloom reached under the counter and produced a giant-sized leatherette album and guided us down to a counter in the rear of the store. At his invitation I flipped the pages. And there was Rosenbloom, in picture after picture, involved in every sexual act imaginable with fair-skinned women.

"Something, huh," he said. "We have a full studio room downstairs where we test all the equipment. I used a time delay on these, absolutely one hundred percent reliable."

He went on with some technical jargon about cameras. I was not listening. Rosenbloom had a tremendous erection in every photograph and I could not figure out how any man could set a camera, pose in front of it with a naked girl, and achieve such full thrust before the shutter clicked. Otherwise, I was not taken with the pictures. Sex, in my view, should be between two people in private, not photographed and shown to strangers.

Not wanting to appear unappreciative of the shared confidence, I leafed through the pages with suitable flattering comments, closed the album, and changed the view by glancing around the store. The salesgirls, four of them, briskly and prettily about their work, suddenly looked familiar. They were Rosenbloom's porno partners!

I turned to him in astonishment. "You mean that these girls just go downstairs and pose with you like this?"

"Sure," he said. "If they want to work here. And a lot of them do. I'm very good to my staff. Listen, it's not an exclusive thing. If you want to go downstairs and make the scene you can have your pick of the girls."

"Thanks, Eugene," I said, laughing. This guy was not even cracking a smile. "But, first of all, it doesn't turn me on, and, second of all, I'll wind up in one of your photo albums."

We were about to leave the store when my curiosity could not be contained any longer. "I got to know," I said, "how a Japanese guy acquired a Jewish name like Eugene Rosenbloom."

"In this line of business," he said, waving his briar pipe at the electrical appliances, "you're better off with a Jewish name. The best stores on Pitkin are run by Jews. Anyway, isn't Sammy Davis Jewish?"

"Sammy Davis is an entertainer," I said.

"Well, you saw the book. Eugene Rosenbloom is an entertainer too."

Chapter 5

The Rockaway
Caper

BY the spring of 1968 I felt well-blooded in the arson wars and I went into a fresh assignment in Rockaway almost as a vacation—a euphoria that lasted less than two hours.

The infamous Charlie Brewer and I drove out of Manhattan via the Brooklyn-Battery Tunnel, electing a fast ride on the Gowanus, around the horn of Brooklyn. We sped past the huge warehouses of Bush Terminal and came to the smell and sight of the harbor at Owls Head. The house lights on the Staten Island hills twinkled across the water, and the parking bays all along the Gowanus were packed with the cars of young lovers. On the stone archways of the overpass bridges, our lights picked out the graffiti of the day, "I love you Sylvia" and "Johnny loves Sue, 1967." Beyond the great new towers of the Verranzano Bridge, by Fort Hamilton, freighters from a dozen foreign countries rode at anchor and I wondered if those sailors had ever seen a more beautiful sight than the necklace of lights strung four thousand feet across the water to their north. Past Gravesend Bay we swooped on to the straightaway, east across Coney Island,

and on through Sheepshead Bay to the Flatbush junction. Bright yellow-white mercury lights overhead showed a clear road by Floyd Bennett Field to the Breezy Point Bridge, where men and boys with lanterns fished for hake. The land was sandy here in the South Rockaways, with open patches of reed and stunted pine and then blocks of Spanish-tiled houses. Beach Channel Drive took us to the blonde brick cooperative apartment houses of Rockaway Beach Boulevard and when we hit the clutter of stuccos and old summer bungalows and the sad neglected hulks of the wooden Queen Annes, and we stopped for coffee at a diner by the Playland, I showed Brewer my watch. We had been on the road less than thirty minutes.

The ocean front blocks on the central strip of the Rockaway split, which divides the Atlantic from Jamaica Bay, were crowded with bungalows that had once been the seaside escapes of affluent Manhattanites. The elegance of parasols and straw boaters and bandstands had long faded into the past, however, and most of the little houses had fallen into disrepair. Poorer families from Brooklyn still rented them in the summer, and there were some year-rounders who resolutely kept their paintwork gleaming and their shrubbery flourishing, but the district had a weary, stained, beaten look about it. Predictably, the degeneration attracted the degenerates, and we were there because they were trying to burn the place down—haphazardly, and in different ways, but building-by-building they were torching Rockaway. Our instruction was to become part of the scenery, which was much easier than in a ghetto like Brownsville, and remove the worst of the firebugs. For this, on a warm May night, we were dressed in teeshirts, cut-off dungarees, and sneakers without socks.

After the coffee we drove down the bumpy roads to the foot of 86th Street, which was a sandy, dimly-lit thoroughfare to the beach and boardwalk between two rows of about forty deserted cottages. It could not pass an automobile. The men of local Engine Company 256 had answered five alarms on this block in the past week, and it seemed a good place to begin.

We were on foot, our car at the corner, when we spotted two figures running into one of the buildings some two hundred feet down. Our eyes met in mute acknowledgement of the sighting and we kept on, our guns drawn, to what we estimated was the bungalow that had been entered. Brewer waved me around to the back while he proceeded up the five wooden steps to the screened front-door. The cottage was barely twenty feet wide, the rooms going back, shotgun fashion.

Sand poured into my sneakers as I crept along the side of the house. I stepped over a three-foot picket erosion fence and around and up the back steps to the porch. The screen door was partly ajar; the main door opened at the soundless twist of the handle. I stood in the doorway letting my eyes adjust to a deeper darkness, then proceeded through the kitchen area. There was a short hallway beyond to another room. One step inside this room and my eye caught the shadow of a man to my left front. My blood ran cold. He was pointing a gun at me.

"Halt! Police!"

I sprang back as I shouted and opened up with five rounds from my short-barrel .38 pistol.

A deafening shattering of glass accompanied the few seconds of gunfire and even above this I heard Charlie Brewer screaming from the front room.

"John! John! Is that you? Goddam it to hell! What happened!"

Brewer was running toward me, crashing into furniture, cursing something awful.

"It's okay, Charlie. A man with a gun. I got him. I must have." I was shaking like a feather.

Brewer snapped on his big flashlight, first illuminating me and then the source of the breaking glass in front. A closet door hung wide open, its full-length mirror blown to pieces.

"Jesus, Charlie," I said in awe. "I shot my own reflection."

I tittered weakly, and he joined in until we were roaring. The hysteria was a relief reaction to a violent fright. We were aware too that it could have been Brewer I faced in that spooky house, Brewer that had been gunned down instead of a fly-specked mirror. We had been foolish. I should have waited on the porch, guarding the exit, while Brewer searched the rooms from the front.

Nervous from the incident, we did not leave each other's side while we swept the next three cottages. The intruders could have run for the hills at the sound of the shots or they could be lying doggo, thinking we had given up. At the fourth house, only one lot back from the beach, we came up empty right through to the kitchen. A small room had been added to the ocean side of the kitchen; the door to it was closed. Brewer took one side, I took the other. Brewer lifted his foot and kicked hard, flat-soled, at the timber by the handle. We took one rapid step into the room as the door snapped open. Our guns were extended in our right hands and supported at the heel by the left.

"Freeze!"

We saw two pairs of bare feet, entwined at the foot of a double bed, then the feet jumped apart and Brewer's flashlight shone on the stricken faces of a teenage boy and girl. They were naked; we had caught them in the act of intercourse.

The kids threw on their clothes and when we took them out onto the street the girl was crying and the boy was pleading for a break because his uncle was a cop.

"Take it easy, both of you," I said. "We'll end the matter here. But never, never again go down in one of these bungalows. There's been a series of fires here and we're staking out the area. You could get hurt."

As they turned to leave I said, "By the way, didn't you hear shots a while back?"

"Is that what it was," said the boy. "We thought we

were hearing fireworks."

Brewer grinned after their retreating forms.

"Fireworks, eh. They must have really been going at it if they—"

"Let it go, Charlie, let it alone," I cut in wearily.

Man, had we screwed up that night. And the worst was still to come for me.

In succeeding block-by-block searches of untenanted cottages and larger summer places we rounded up twenty-odd junkies, alcoholics, and just plain loafers, and had them booked for trespassing. Word spread through the subculture that fire marshals had launched a blitz and the incidence of fires dropped off sharply.

But there was one persistent arsonist who took our presence in Rockaway as a challenge. His fires increased. We knew it was the same person because he did not vary his technique—always a mattress set afire at the rear of a building or under the stilt-raised back porch. The houses were tinder-dry and this kindling, while it started slowly, was sufficient to make a bonfire of them. Close to one hundred buildings had been razed by this madman. As the weeks went by we had the impression that he was watching our every move. If we were busy at 56th Street he would torch a place on 66th, even in the daytime. It was his sport to make us appear clods. Several times we rolled in on fresh fires with the rigs to find a taunt scratched into the sand: "Fuck you fire marshals."

We knew, or thought we knew, the identity of our tormentor. Firemen had reported a big Irish-looking youth with long, fair curly hair near the scene of many fires. He would stand off to one side, arms folded, leaning against a telegraph pole, while he watched the play of hoses. Terence Kelly was his name, we found, nineteen years old, unemployed. He lived with his mother in a weather-scarred stucco house by the elevated railway west of the summer cottage area. She called down the wrath of God on Terence if he so much as looked at a girl in the street.

She decided that harlots were waiting on every street-
corner now to steal her son. When we questioned him
about the fires he simply said: "Leave me the fuck alone."
We had to do just that. As he was fully aware, we had
nothing on him. The notion entered our heads in frustrated moments that we should flake him, that is, falsely
swear that we had witnessed him setting a fire. Brewer
said we could justify it as a service to society. I could not
try it on. To me flaking was the most heinous violation of
a cop's honor.

During the summer Rockaway filled up with vaca-
tioners, and the hired insurance torches and the firebugs,
including Terence Kelly, went into recess. Brewer and I
were involved in separate patrols elsewhere in the city
and when I returned to Rockaway in September it was
with a new partner, John Connell, son of the renowned
Dr. John Connell, a fearless and much respected man
who had served as a medical officer to a generation of
firemen. Young Connell, in his early twenties, was a law
student as well as a marshal. A tall, athletic figure,
trimmed black hair, and a nicely cut face gave him the air
of a tennis coach from a New England college. His idea of
working street clothes for the Rockaway tour was Ber-
muda shorts, Indian sandals, and a golf shirt with a little
red umbrella on the pocket.
With the flaring again of the fires after Labor Day an
early mission was to examine as possible arsonists a gang
of about twenty hippies that loitered day and night under
the boardwalk. I call them hippies, though they ranged in
age from eighteen to thirty-five, had no interest in pot or
politics, and were in a constant state of semi-intoxication
from Thunderbird wine.
Unshaven, unkempt, and carrying a bottle of Thunderbird, I was readily accepted. In fact, I went all the way,
drank the wine right along with them and slept out on the
beach. A dirty vest covered the bulge of a small transmitter radio taped to my chest. I kept it on in case my cover

was broken and Connell needed to come to the rescue. My partner bicycled up and down the boardwalk with a radio in the pocket of his shorts and the aerial running up inside the back of his shirt. It seemed a perfect ploy to Connell—a man getting his exercise. Trouble was that he really crouched over that bike, feet pedaling like the devil. The way that his radio and aerial stood out against the stretched material of his shorts and shirt he might as well have strapped a flashing red light to his head. Everybody made him for a cop.

For three days and nights we made this scene. Connell got his waist down to thirty-two, I got a monumental hangover, and the hippies were left to their dissipation. Arson was not their bag.

A series of mattress fires in the cottages in the far east blocks of the beach made us suspect that Terence Kelly had renewed his campaign, and we decided next to concentrate our watch in this area.

For the preliminary patrol Connell drove the car right up onto the boardwalk, which seemed like a fine idea until fog came billowing in from the sea. We were inching along, our headlights penetrating less than twenty feet into the gloom, when the ambling figure of a short, fat man loomed up in front. He was swinging a kid's baseball bat at his side.

"Hey, pal, what are you doing here on a night like this?" I called as we drew abreast.

"The fog brings me out," he said. His voice sounded hollow and he spoke very slowly. I assumed he identified us as police officers and was putting us on.

"Why do you like the fog?" I said lightly.

"It makes me feel secure." His tone was unaltered.

I touched Connell's arm to signal him to turn off the motor and babbled on with the stranger about how warm moist air from the land was mixing with cooler air from the ocean and creating the fog. It was a pity that it was not raining, because rain extinguished the fires that were being set in the area.

"Do you know about the fires?" I finally got to the point.

"Yes, many fires here." The man's face was thrust close to my open window. He was around thirty years old, matted hair, an acne-pitted face.

"Do you ever go into these bungalows?"

"There is one I like," he said. "It has magazines with pictures of girls." He broke into a high-pitched giggle, the opposite register to his speaking voice, and pumped the baseball bat up and down in his hands.

I removed my gun from the hip holster and held it across my lap.

"Do you ever see fires while you're in there looking at the pictures?"

"Sometimes," He paused as if trying to remember something. "Yes, I used to make my own fires. Out on the beach. I used to burn Christmas trees."

"What about now? When you're looking at the girls? Do you set mattresses alight?"

They were careful questions. Marshals had pursued a host of demented creatures, male and female, who lit fires to facilitate masturbation. Apparently it was the only way they could get off.

"Once in a while I have to burn the mattresses," replied the fat man. Again he let go with that girlish laugh.

He climbed into the back of the car with me, meek as a lamb, while Connell found his way off the boardwalk and across to the 101 Precinct. We booked him for arson, but were pleased next day in Queens Criminal Court when the judge recommended psychiatric treatment, and the episode ended there.

There was an article that week in one of those local, give-away newspapers about the concerted effort by the Fire Department to apprehend the Rockaway arsonists. Recognition was given to fire marshals Barracato and Connell for capturing the "mattress maniac" of the east shore. Fine and dandy, except that Terence Kelly must have seen the piece and set five separate fires late in the night of publication day—all mattresses, all in the east beach area. We inspected his work the following morning. Printed in the sand by the fifth house put to the match was the familiar calling card: "Fuck you fire marshals."

Kelly, if it was Kelly, must have been disappointed with the result of this fifth fire. The mattress had been hauled out into the back kitchen and lit, but the doors and windows of the house—a two-story, white shingle place—had been broken or ripped out by vandals and the timbers were damp from rain. The fire had barely eaten up the kitchen wall before the rigs arrived and knocked it down.

The arrogance of this particular arsonist convinced us that he would try again. At nightfall, we left the car two blocks away and crept through backyards to the white house with the idea of searching it and then staking it out. A street light directly outside showed our way up the front steps and on through the gaping entrance to the parlor. Sitting on the floor, against the wall, we saw two crouching men.

"Freeze!"

The word was said, not shouted. I was irritated. Kelly was not a man to have an accomplice.

We had stumbled across two junkies, one with a hypodermic needle and caps, though no heroin. There was a possibility that they were also arsonists, so we played safe and took them to the precinct squad room for booking as trespassers and for questioning.

I had rolled in the paper to commence typing the arrest forms on the oldest typewriter in New York City when one of the men got up and announced: "Officer, I gotta take a leak."

I looked at him with something I fear was close to loathing.

"The way you smell I thought you just went ahead in your pants."

Like the other man, he was white, about twenty-three, a three-day beard, thin as spaghetti, and he stunk to high heaven—a typical skel, as we called them, skel for skeleton.

"Wait until I type this."

"No," said the skel, "I gotta piss bad."

With a sigh heard around the borough I left Connell to watch the second junkie while I escorted the pisser to the two-tone green bathroom in the passageway outside the squad room.

As soon as the door squished shut on the overhead air lock the man turned around and looked me straight in the eye. "You guys are going to blow my whole plant. I'm a cop, a narcotics officer."

"Oh shit." My brain went numb. Rockaway was too complicated.

"I carry no proof, you understand, but go ahead and call my sergeant at the one-fourteenth. Ben Durocher. That's me. He'll verify. But whatever you do, when we go outside, don't give me away. I been a month on this goddam plant."

"How in God's name can I up and cut you loose now?"

"I don't know," said the narco man. "Just do it right, that's all."

I returned the officer to the squad room and checked him out at the desk telephone downstairs. "It's very important," said the sergeant, "that you don't arouse the other man's suspicions."

Upstairs, I briefed Connell in the hallway and we went back into the squad room like goons.

"We'll finish the typing later," I said. "My partner and I got a tip that you two tried to torch that house last night. The one we caught you in. We want an admission, tonight, now."

The second junkie, who was nodding and deathly pale, told us we were nuts. The undercover cop started laughing.

"Funny, huh," said Connell. "What do you say to a little dental work?"

He picked up the black handpiece from the cradle of the telephone on the table, wrenched the cop's head back by the hair, and poised his unique weapon as if to smash it full in the cop's face. Connell's eyes blazed; he was capable of going all the way.

The second junkie was on his feet then, pleading. "Stop it, for crissakes. We didn't set no fires."

"Bullshit," I said, gratefully watching Connell release his grip on the telephone. "We got you cold. John, you take the collar."

"Not me, buddy," said Connell. I loved him for it. "I got things to do. No time for court."

"I sure as hell can't go." It was my turn now. "My kid and I are going to Shea."

I slumped down on a chair, thinking I should be on Broadway.

"This is your lucky night," I said to the junkies. "We're going to give you a break."

Both of them came forward in their seats, expectantly.

"Get lost," I said.

(Three years later, at senior detective school at the Police Academy, a trim young cop came up to me in the library one day. "Remember me?" he asked. "No, I'm sorry, I don't," I said. "Rockaway '68," he said. "You and your partner, the college-looking brawler, went into the best ad-lib routine I have ever seen to keep my cover good. I'm Ben Durocher.")

Terence Kelly was to put Connell and me through one more humiliation before we got him.

The city was acquiring a string of abandoned Queen Anne homes along Beach 101st Street and demolishing them to clear way for a new development of high-rise apartment houses. Several remained tinned up, awaiting the wrecker's hammer, and night after night engine companies responded to alarms there. These buildings were three-story, upwards of eighteen rooms. A fire turned them into flaming beacons that could be seen for miles. Commercial airline pilots on the Jamaica Bay approach to

Kennedy Airport joked that if their own ground radio directional signals went on the fritz they would home in on a Rockaway fire.

Around eleven-thirty on a chillingly cold November night we sneaked into the largest of these Queen Annes, second in from the boulevard corner, and prepared to hide there until dawn. The house was yet untouched, a tempting target for a firebug. Various pieces of valueless furniture, including beds and old mattresses, had been left around, and our reckoning was that Terence Kelly himself might give it a toss.

We huddled down on the floor of a small room off the second floor hallway in our long-johns, heavy winter trousers, and topcoats, prepared to sit it out. Except for the glow of our cigarettes it was pitch black and the cold was so biting that our joints seemed to freeze in place. We could not move about for fear an intruder downstairs would hear us. In our stupidity we did not think for a moment that we would not hear him. This, despite the howling of the wind outside and the heavier gusts that made the building vibrate and set the tin sheets on the windows to banging like cymbals.

At 3 A.M. I looked across to Connell. "Say, John, do you smell smoke?" His nose made audible sniffing sounds. "Holy Christ, yeah!"

We hit the door together, flinging it open to the hallway, and caught a blast of heat in the face. Smoke was pouring up the stairs, an orange glow filling the place from flames somewhere in the back of the house. Our trap had been sprung on us. There was no safe escape because of the seals on the second floor windows. Connell faced back to our hiding room, took one deep breath, and plunged down the stairs, crouching low, brushing the wall, with me charging behind. Our fists pressed hard against our mouths. The heat, inhaled, could sear our lungs, swelling them, cutting off breath. Connell took the last steps in one jump, sprinted through the thick smoke in the hall, and flung himself through the open front door.

Right on his heels, I felt that whiff of cold air and dived for it—tumbling down the porch steps and landing on Connell at the bottom. I rolled on my back, gasping. "Thank God," I said, looking up at the sky.

Connell shouted in the alarm to the Queens dispatcher on his walkie-talkie and then we searched the area around the house for the arsonist. Nothing. The rigs were on the scene within four minutes, extinguishing the blaze before it had spread beyond the kitchen and two rear ground floor bedrooms. In the soggy, charred ruins at the back we found a burnt-out mattress.

"Terence Kelly," I said to Connell next day, "is running us ragged because he's downstairs when we're upstairs and he's on one block when we're on the other. Okay, let's accept that we'll never be lucky enough to be in the same place at the same time. Therefore, it we want to catch him, and we must, we have to devise an alternate method of proof against him, other than eyeballing him in the act. We can set up hidden cameras with trip wires, but that would cost a fortune and they'd burn in the fire anyway. Fingerprints are out because the smoke fills them in. There is another way, though, and he'd never suspect it because he'd never see it—ultraviolet. We could spray ultraviolet powder on all the door knobs and threshholds in one particular area, wait for a fire, and then grab him quick. If his hands and the soles of his feet, or sneakers, whatever, show color under the ultraviolet light, we got him."

Connell thought the plan was worth a try and so did Vincent Canty when we drove up to Manhattan and put it to him. Canty has been chewing us out over the mattress fires, and he had caught a blast from the Rockaway cops over Connell's bicycling act.

Our division had several ultraviolet kits stored in the locker room. They came in a black leather case about the size of a doctor's bag and each contained a battery-operated lantern light, two flashlights, perhaps a dozen aerosol cans of different colored sprays, and an equal number of crayons. Marshals sprayed fire boxes with ultraviolet at locations where false alarms were a problem. A check with the light on the hands of the kids standing by watching the engines usually turned up the culprit. Ultraviolet was also employed to mark money in arson extortion cases. A hood would threaten a storekeeper with fire and demand, say, a thousand dollars. Marshals supplied the tainted money, instructed the storekeeper to make the pay-out, arrested the extortionist, and proved his skulduggery with the beam of the trusty ultraviolet lantern.

By a stroke of fortune, our first in Rockaway, we had good information on where to begin the experiment.

A red-haired woman named Margot—Margot the Harlot we called her—ran a modest, one-woman, red-light house on 49th Street, an oasis of fun in a thicket of forlorn and empty bungalows. She caused no trouble and the law regarded her kindly because she kept a watch on the area and reported anything suspicious. We had often spoken to her about Terence Kelly, whom she knew by sight, and the afternoon after we picked up the ultraviolet kit we made a routine call on her. Kelly, she said, had passed by her bungalow twice that morning and she had seen him at the boulevard corner later when she went off for groceries. It seemed to us he was making a reconnaissance.

For the rest of the day and into the evening Connell and I visited every house in the four-block area between 47th and 51st. At one place we would put red on the door knob and yellow on the threshhold, change the sequence on the next, make it a red knob and a blue threshhold at the third house, reverse that, introduce the yellow again, alter the patterns with green and silver and mauve, and so it went on, every possible combination. It was slow work initially. The spray cans hissed at the touch of the button, but, of course, the ultraviolet was invisible. We kept shining the light to make sure the blue was really blue. After a few houses we decided to trust the labels. Altogether we marked 164 houses and had notes of every color coding by address. These notes we registered at the 101 Precinct. There could be no mistake. If one of these bungalows was torched and we found Kelly anywhere in the vicinity his hands and feet should positively prove his guilt or innocence.

At ten o'clock we parked the car two blocks west of the marked zone and separated for a foot patrol, agreeing in advance not to enter any of the buildings. We each carried a walkie-talkie for communication.

The streets here were actually thin ribbons of pavement between wooden slat walks. Then it was only ten feet or so of sand to the fronts of the stilt houses on either side. Cornell took 51st Street and I went down the middle of 50th, walking towards the ocean. The night was clear and cold, a high crescent moon showing faint visibility beyond the pools of light from the widely spaced electricity poles. It was a promising night for arsonists and we were keyed for the hunt.

Midway down 50th I saw the figure of a man dashing from the line of houses at the end of the block. He turned down the pavement, directly under the last streetlight, sprinted to the boardwalk along the beach, and went for his life—going towards the Forties.

I was off and running the instant I spotted him, holding the button of the walkie-talkie open to alert Connell.

The moment the intruder jumped to the boardwalk I yelled, "Halt! Police!" Then, without breaking stride, I fired two shots into the air.

Connell, meanwhile, came dodging through the house yards and vaulting the sand fences like a cross-country champion. He caught up with me on the boardwalk. I was gasping for breath and slapping the side of my leg with the flat of the gun in exasperation. My quarry had vanished.

The words came out jerkily. "Saw him, John. Big blonde guy. Dark clothes. Looked like Terence Kelly. Think he set a fire back on 50th."

"I don't see a fire," said Connell.

"Takes a while to catch." My mind was working quickly. If we took time to ascertain whether indeed a fire had been set we would lose whatever slim chance we had left of finding the arsonist.

"Let's get the car," I said.

Within three minutes we were cruising down Rockaway Beach Boulevard on the north side of the bungalow blocks, pausing at each corner and looking down the streets. I was aching with disappointment. We might never get this close again.

Between 42nd and 41st our headlights illuminated a man striding along the right sidewalk. He wore a brown leather jacket, dark blue denim pants, and he had a mop of fair curls. Terence Kelly. Connell coasted the car a few yards ahead. We stopped against the curb and called to Kelly.

"What the fuck do you guys want now?" He recognized the car straight off.

"Terence," I said evenly, "get in the back of the car before I kick your goddamned ass in."

"You got nothing on me," he said, not moving.

"Get in!" I was out on the walk now, yanking open the door.

The words were hardly out when we heard the fire sirens. A mile to the west the sky was aglow from a runaway blaze in one of the beach cottages.

"Get in!"

I sat in the back with Kelly. Connell turned the car and returned to the foot of 50th, where the rigs were assembling. The fire was there all right. Connell went off to get the bungalow number, taking with him an ultraviolet flashlight to double-check the colors on the door knob and threshhold against our registered list.

I was watching the firemen hooking up to the hydrants on the boulevard when Kelly put his arm through the open window and slipped the back door lock from the outside.

"I'm getting out of here," he shouted.

I was onto him before one leg had reached the walk, wrestling him around the shoulders to force him down on the seat. He was six feet tall, two hundred pounds, compared to my five-nine and one-fifty. His hand thrust under my chin, damn near snapping my neck.

At this precise moment a fireman stuck his head in the window. "Hey, move this car. You're in front of a hydrant."

"Forget the hydrant," I said, choking. "This is the firesetter."

The two of us had Kelly subdued and cuffed by the time Connell returned. John was very excited.

"Perfect," he said, jumping behind the wheel. "Beautiful. Colors sharp as a rainbow."

The cops at the 101 Precinct were as curious as cats when we rolled up with Kelly. They had been heckling us for weeks about our failure to bring him to account, and I was sure that they felt the complicated business with the ultraviolet was merely a showy exercise to prove we were still trying.

"Douse the lights, lieutenant," I instructed the senior officer, a stocky man with a beer belly, when we escorted Kelly to the squad room.

Connell flicked the switch on the ultraviolet lantern and my heart leaped when the beam showed a strawberry red glow on the palm of Kelly's right hand.

"There it is." Connell's voice was elated. "Same color as we put on the knob of number nine 50th Street, the place that burned down."

Kelly pushed away the light. "What kind of a frame is this," he snapped. "I could have got that stuff on me anywhere."

"Relax, Terence, we haven't finished," I said quietly.

"The hell you haven't." The husky blonde lad was on his feet, his pale face contorted in rage and bewilderment. He spat his words. "You fuckers, you rotten, stinking fuckers!"

"We don't need that," I said sharply, talking him down.

Kelly's eyes clouded over. He was more docile as Connell ran the light over him. Various color smudges showed on his jacket, where I had handled him with my spray-marked palms, but vitally, there were pronounced red streaks in his hair, red on the shell of his left ear, and red on the fly of his jeans.

"Open your jeans, Terence," I said.

"No way."

"We'll have to remove them for you."

He unzipped to reveal red on the front of his jockey shorts.

"Your penis, Terence."

Reluctantly he fished it out, and we stood there, two fire marshals and five cops, staring at the strawberry red glow on his member. After the chase he must have stopped for a leak.

Nobody said a word as I asked him to zip up again and give us his sneakers. He was getting intrigued with this strange routine himself. Sure enough, the soles gave off a blue light, matching the door sill color on the razed house.

"Terence Kelly," I said, "we're booking you for arson in the third degree."

The case went all the way to the Queens Supreme Court and in contrast to the Cora Sue Fricker trial fiasco, the defense attorney's attempts to ridicule my evidence rebounded against the defendant.

Our use of the ultraviolet, and its discovery on Kelly, was recorded in general terms before the lawyer, a thorough, middle-aged type with rimless glasses, sought to discount its importance by getting to specifics.

"Marshal," he said, "when you and your partner color-coded these buildings with this so-called ultraviolet material, did you wear gloves?"

"No, sir, there was no need for it."

"Answer without ad-libbing, marshal. All right, then, is it reasonable to believe that you had various colors on your hands?"

"Yes, sir, many colors. We did four square blocks of buildings."

The attorney seemed pleased. "When did you arrest my client?"

"About twenty minutes after we chased this figure down the boardwalk," I said agreeably.

"Did you at any time place your hands on my client?"

"Yes, sir, I did. I handled him about the shoulders and arms."

"Any other places?"

"To the best of my recollection—no."

The questioning then probed the examination at the precinct, the finding of ultraviolet colors on Kelly's person.

"In essence," said the lawyer, "did you see colors on my client where, in fact, you touched him?"

"Yes, sir."

I could have jumped in there with the full result of the color scan but the lawyer might have cut me short. I waited to be led.

"Where else did you see colors?" The lawyer gave it the tone of an unimportant mop-up question, but it was the one I wanted.

"He had red ultraviolet color on his hands, his hair, his ear, and also on his penis."

The court hushed completely at the sound of the word—for a split second no coughing or scraping of feet, not a peep. Then a ripple of laughter went through the gray room, and it lasted until the judge banged his gavel for silence.

The defense lawyer made a grimace and announced no further questions. He had tracked the minds of the jurors onto the notion that I had put the colors on Kelly myself. I might have touched his hair and ear, but never his penis.

We did not even get to the evidence of the marked sneaker soles; it was superfluous. Terence Kelly was found guilty as charged and sent to prison.

Chapter 6

"R.F.K. Is Gonna Burn"

THE marathon Rockaway pursuit had been interrupted by a variety of other cases, but none more nerve-wracking than the hunt for a psychopath who was plotting to make a pyre of Robert F. Kennedy's bier in St. Patrick's Cathedral.

The body of the murdered Senator was flown to New York on the evening of June 6, 1968, for the public viewing, and somewhere in the city was a maniac intent on making it a public cremation.

Two days earlier, Connell and I had responded to a suspicious fire in a tavern on Clarendon Road in Flatbush. The back rooms of the bar had been gutted. We inspected the area, wall by wall, until we found the tell-tale cone of scorching that marked the origin point of the fire on the plaster in the bathroom. Definitive lines across the cone marked the rising heat waves. At the foot of the triangle of burn, a yard out on the linoleum floor, were burnt fragments of a plastic valise, a man's chino pants, and plaid jacket. The deeper charring of the linoleum and floor timbers underneath these remnants confirmed that the fire started here, almost certainly a bag of clothes set alight with a match. There were no electrical wires or outlets, and no gas jets nearby, to indicate an accident. And the floor charring was not severe enough for a gasoline blaze. Accelerant fires could sear through five or six inches of wood in heavy floor timbers in a matter of minutes. A regular fire ate into such wood at the rate of about one inch every hour. Barely a half inch had been burned away here.

Among the remnants of clothing, we found a piece of a hard plastic badge, like an identification tag. We could make out a number, 147, and a location, Glassboro, New Jersey. A fast trip to Glassboro got us the information that this was a migrant farm worker's registration badge—and No. 147 had been issued to a man named Fidel Zavos late in May.

The name was familiar. The previous February I had arrested a Fidel Zavos for torching two churches in lower Flatbush. He had stood outside the second one, St. Matthew's, watching the flames light up the stained glass windows by the altar, and loudly proclaimed that his mother was the "queen of space" and his father "the god of gods." His parents were both dead, as it turned out, but Zavos said they had come to him in his dreams and commanded that he burn down all the churches on earth and turn people away from the "false" God of the Bible. Zavos went straight from court to the Stoneville psychiatric hospital near Hicksville, Long Island, and I had assumed he would be there, under tight security, for a long, long time. Stoneville confirmed that Zavos had walked out of a hospital recreation room four weeks earlier and vanished.

We knew he had gone to Jersey for a brief period, but his fiery trail in a Flatbush bar indicated that he was back in his home territory and probably ready to strike against the churches. We returned to Flatbush on the afternoon of June 6 and began checking with local pastors to see if they had seen a wild-eyed Latin type with a pencil-thin mustache loitering about their churches. Some time during the night we had a message to contact Stoneville. I called from the Brooklyn dispatcher's office. One of the orderlies at the hospital had had an hysterical telephone call from Zavos in which he shrieked that the American people were now making a god of Robert Kennedy. His mother, the "queen of space," had ordered him to go to St. Patrick's Cathedral and burn Kennedy's body in the fires of hell.

"We gotta problem, John, an almighty problem," I told Connell when I hung up the phone. "Zavos has threatened to burn Robert Kennedy's body when it lies in state at St. Patrick's, and he's capable of carrying it through."

My mind would not dwell on the shock such a bestial act, or even an attempt at it, would send through an already grieving and shattered nation.

"We gotta find him, but where do we start looking?"

"It has to be in Manhattan or a subway ride away," said Connell.

"Right. Manhattan, or his familiar stamping ground of Flatbush or Canarsie."

"He doesn't have any money, as far as we know. Probably no regular place to stay."

"Right. That means a very cheap hotel or a flophouse."

Connell was on his feet. "Let's go."

For the rest of the night we visited every flop we could find in lower Brooklyn and lower Manhattan, including the fifty-cent-a-night joints off the Bowery. We had printed up dozens of pictures of Zavos for distribution to clergymen, and these we left with the night clerks with instructions to call our division headquarters if they spotted him. None could remember seeing him that particular night, but the clerks in these lousy hell-holes did not look at faces as a rule—just the silver in the palsied palm.

At the Secret Service command post at St. Patrick's next morning, as the body of the assassinated Senator was being prepared for the viewing, we told of the threat and distributed more pictures. Agents agreed that we should join the detail assigned to watching the public lines forming outside, and for identity we were given tiny, round, enamel lapel buttons. The buttons were tricolored, the color of the day, red, to be worn topmost. We were further instructed not to mention Zavos's threat to any members of the Kennedy family.

Through that long, sad day at the cathedral, we waited and watched for Zavos. People were admitted for a time through the doors at the side of the central entrance, where they could be more easily studied by the Secret Service, but the Kennedy family asked that the main doors be opened. I was moved by the composure of the Kennedys; they were splendid. Jacqueline Kennedy thought to speak to several of the agents, thanking them for being there.

Of Zavos, however, there was not a sign, and around midnight I went home for sleep. The Senator's body was to be moved to Washington late the following afternoon. Soon the danger would be over.

At 4 A.M. the telephone rang by my bedside. It was a duty marshal from the bunkroom at headquarters. The desk manager at the Jackson Hotel, in the Bowery, had called to say that Zavos had just checked in.

I alerted John Connell at his home in Queens and we rendezvoused thirty minutes later outside the Jackson, a narrow, five-story brownstone hotel with a flickering green neon outside and a strip of threadbare carpet that started at the threshhold and led through glass doors and down a hallway to the registration desk.

The clerk lifted a pink, scaling face at our approach.

"Your party's in cubicle twelve, second floor. You want the light on up there?"

"No light. And you stay here. Is the cubicle locked?"

"Hook-and-eye catch, from the inside."

We climbed the stairs to a huge open ward room. Fifty beds, at least, were pushed close together down either side of the wall and most were humped with sleeping men, and, probably, women too. This was skid row at rest.

"Jesus!" Connell grimaced and pinched his thumb and forefinger over his nose. The stench of sour wine and sweat and urine was overpowering.

We walked down the aisle between the beds to the

cubicles, like toilet stalls, at the far end of the room. For the privacy of a bed in one of the stalls the tab was hiked twenty-five cents.

Twelve was marked with big house numbers nailed on a plywood door. It was open at the top, the walls about seven feet high.

"I'll go over the wall," I whispered to Connell. "This guy carries a knife. I remember from last February. If we bust in there he might have time to use it."

I gripped the top of the wall beam, threw a leg up, and was quickly over the other side. I flipped the door lock and then grabbed Zavos under the shoulders and threw him bodily to the floor.

"Freeze!" Connell had his gun pointed at Zavos's temple.

My right hand went under the pillow. Nothing. I slithered it under the mattress and felt the hard butt of a knife. As I withdrew it I felt a sting on the top of my hand in back of the first knuckle.

"May I know what is going on?" Zavos's voice was very English, very formal.

"No, you may not, you sonofabitch," said Connell.

We handed him his clothes piece by piece from the chair at the foot of the bed, then we handcuffed him and took him out past the sleeping derelicts. Not one of them had stirred through the whole proceeding.

"What in Christ's name do you think they are dreaming about?" Connell muttered.

"Thunderbird wine," I said. My knuckle itched and I was sucking on it.

In the car we searched Zavos's pockets and found the key to a locker at Pennsylvania Station.

"Is this where you keep the shit?" asked Connell.

"I do not make shit, I make fires," Zavos replied. His tone was calm but he had a silver glitter to his eyes that made me shiver.

He pulled on my shoulder. "Thank you for catching me, marshal, once again. I was going to burn Bobby Kennedy.

The 'queen of space' told me to burn him and I said I could not. She said I must because my father ordered it."

"Is the shit at Penn?" Connell had no respect for the insane.

"Oh, yes."

We found a brown paper shopping bag in the locker containing a gallon can of paint thinners, plus two quart bottles half full of the same inflammatory stuff.

"I stood in line at the cathedral with my shopping bag today," Zavos explained, "but it was moving too slow, and I decided to leave and go back this morning."

After we booked him at the precinct near Penn Station I called the Secret Service and told them the maniac was in custody. I was assured that he would have had no chance of getting into the cathedral carrying a shopping bag full of paint thinners in bottles and cans. But I wondered about that. Only marshals are trained to think of fire as a weapon. Numerous people buy paint thinners for cleaning up paint smudges at home. It is a harmless liquid unless put to the match. Would the agent at the door when Zavos arrived have made a diabolical connection?

I had trouble typing my report on the case because my right hand was swelling up like a melon and turning purple. Whatever bit me in Zavos's cubicle must have been a cross between a roach and a centipede. I carry the scar to this day. A physician available to us at the Municipal Building said I had blood poisoning. He drained the infection and swathed my hand in bandages.

One of the press information officers at the Fire Department, in issuing a short account, talked to me and, seeing the bandages, assumed I had been wounded by a knife or gunfire. The *Daily News* carried a story and the next day I was instructed to appear at a seminar being conducted at the United Nations by the American Federation of Police. The subject, I believe, was the bombings that were then part of the antiwar résistance in the United States.

I was called to the platform at the seminar and grandly

presented with a plaque honoring me for meritorious conduct.

"And as you can see," said the police captain making the award, "Marshal John Barracato was wounded in the line of duty."

Shamed to silence, I neglected to mention that my wound was a bug bite suffered in a flophouse in the Bowery.

Chapter 7

Ambush!

I WORKED with several different partners after my Rockaway spell. The Fire Department was packing up to move
across Manhattan to its own building at 110 Church Street
and coincident with this, and with the explosive growth of
arson in the city, our division was being structured on
firmer foundation. The A and B pairing arrangement was
crumbling anyway, and we settled on a system under
which six supervising fire marshals each controlled a
group of nine equal-rated men. That meant one swing
man and four two-man teams to work each of the five
boroughs on round-the-clock scheduling. The Brooklyn
car also covered Staten Island, and this continued to be
my first choice. Reason one was that before the Bronx
erupted, Brooklyn was the busiest beat. Reason two I
never confessed to anyone. I wanted to be close by in
case my own house caught fire or some twisted, disgrun-
tled arsonist from my past found out where I lived and
decided to torch my home. We were about to move into a
new ranch in a wooded cul-de-sac off Richmond Road at
the time.

As a further staff reinforcement in the division, the com-
missioners decided to dump the provisional pretext. An
examination, open to all firemen, was held in 1969 to get a
permanent marshal roster. The paper consisted of one

hundred multiple-choice questions. I answered them one morning in a schoolroom in the Bronx. I did not study for the test because I suppose I resented it. A man could be a whiz at test papers and hopeless out on the street. I passed tenth out of the 2,700 who took the examination, but it did not make me a better marshal. Thirty-seven other marshals, some of them superb investigators, flunked and were replaced by new men from the firehouses.

The division also lost its ace. Charlie Brewer had worked forty hours straight investigating a grudge fire in which a fire captain had lost his life. He collared the hoodlums responsible and was on his way home to Flatbush from Brooklyn Criminal Court by subway when he suffered chest pains. He was doubled in agony when the train reached his stop and he staggered three blocks to collapse on the front steps of his home. Physicians at Coney Island Hospital saved his life with open-heart surgery. On the dozens of occasions I visited Brewer in the hospital he did not once express regret that he had not taken the retirement offered him years earlier. His main joy was to relive the adventures we had shared. "You got it, kid," he would say, "the town at your feet."

I felt that I had, too, and it was all that I wanted. My path crossed that of expensive lawyers, high-paid television newsmen, physicians who lived in mansions, top-ranked city employees—yet I would not have traded places with any one of them. They worked the sterile high ground of the city, I worked the rich thicket of the streets. My clothes were taking on a suitable resplendence, and I was acquiring the controlled swagger that is an occupational characteristic of homicide detectives. These guys walk into a crowded all-night diner in the worst part of the city like movie gunfighters entering a saloon in the old West. I had seen it countless times; always the pause at the door as they rake the place with their eyes, then a nod to the owner, and a slow walk to a center table with an almost imperceptible rolling of the shoulders and narrowing of the eyes. They cultivated the air of men who felt astride the city, ready and able to grapple with its darkest forces. Sheer conceit, of course, but role-playing is the stuff of life in the streets of New York. It is a town of actors and the more vivid for it.

The proud survival art of being street-wise can only be learned in one classroom, and only by direct experience. You do develop a sixth sense, but it takes a few years and a lot of hairy confrontations to get to that level. Inevitably, you fall into traps along the way, but if you're lucky, a fear-inspired cunning enables you to scramble free and live to fight another day.

John Connell and I had our humbling following a lumber yard fire in Jackson Heights, Queens. During the night, four black youths scaled the cyclone fence around the yard, collected five jerry-cans of gas left by some fork-lift trucks, and emptied them out over an immense pile of two-by-fours. They tossed a match to start a blaze that destroyed eighty thousand dollars worth of timber. We had no doubts about the identity of the arsonists because residents in adjacent houses had seen them climbing back over the fence. The fire lit up the yard like daylight and a number of people recognized the boys as chronic trouble-makers who lived in the Hiram J. Smith housing development three blocks away.

About four in the afternoon we drove into the complex of blonde brick apartment blocks and zig-zagged through the paved, narrow roads until we got to the Q building, where two of the youths lived. They were together in the second floor apartment of the one called Artie. With surprisingly little jiving around they came down and got into the back seat of the car.

"We'se juveniles," said Artie. "I'se sixteen and Rocky here's fifteen. We cain't tell you nothin' without our folks."

"You're quite right," I said from the front passenger seat. Connell was behind the wheel. "You're gonna cause them a lot of grief. Half the city saw you burn that lumber yard last night."

"That was Bennie's idea," said Artie. "Burn the honky's wood."

"Where is your father, Artie? We'll drive you to him."

"He ain't gonna be happy if you go to his work."

At this point in the conversation, when I was twisted around talking to the boy, Connell tapped me on the shoulder.

"John," he said quietly. "We got company."

At least twenty people, men and women and teenagers, had gathered on the sidewalk at the front of the car and were pointing at us and muttering to each other.

"Quick, outa here," I snapped.

Connell fired the motor and threw the gear into reverse. We were facing up a deadend, too narrow to turn. There was no option but to back up.

We were too slow. A group of teenagers had gone to the rear of the car and were making faces at Artie and Rocky through the back window.

"Jesus, I can't run them down." Connell had reversed a few feet, then stopped.

People were everywhere by now, and more were pouring out of the buildings—a hundred of them, it seemed to me, completely surrounding the car. Four young men in Afros, as tall as Knicks players, curled their fists around the roof rim on my side and started rocking.

"Let them boys go," a woman shrieked.

"Where them boys parents," came another cry.

"Pigs! Honky pigs!"

Everybody was shouting then and the car was being heaved from side to side. I seized hold of the dash for support. In back, I glimpsed the boys sitting close together in the center of the seat.

"Call a 10-13." Connell's face was blanched, but his voice was low. He gripped the wheel and looked straight ahead as if he were driving in deep concentration down a busy expressway. Good man, John, I thought. First rule of the streets is never to show fear. It acts as a goad.

I unhooked the radio handpiece.

"Car 56 to Queens dispatcher. Transmit a 10-13. Our locations is—Where in hell are we, John?"

"I don't know," said Connell impatiently. "Tell them the Hiram J. Smith housing development."

My eye caught a big man in a stained blue windbreaker pushing to the front of the mob on my side. He whipped out his penis and started peeing on the fender.

"Queens. We're in the Hiram J. Smith development. I can't tell you exactly where. By the Q building. Just tell them we'll blow our siren as soon as we hear theirs."

I replaced the handpiece and rolled my window down a fraction to be heard.

"Hey," I called. "You. The big buck. Put your thing back in your pants. There's women and children here. You're calling us pigs and fuckers and you're out here waving your sword around like it's a piece of jewelry."

"How'd you like to bite on it, pig," the big man snarled.

"Bite on it," I said loudly. "I'd like to shoot the goddamned thing off."

A fat glob of spittle, a really revolting lunger, splattered on my window. I rolled up the glass as the lungers flew thick and fast, over the windshield and hood, and the rocking commenced again, more violently than before.

We heard the sirens then and Connell sounded our own to home them in. I saw the mob scattering from the edges, disappearing back into the buildings. But many of the young people were not moving. This could get ugly.

I reached over and slipped the trick catch on the back door.

"Okay, kids. Out!"

Artie and Rocky were off like rabbits as the blue police cars roared into the roadway, one after the other, eight or ten of them, coming like the cavalry. The crowd had mostly gone, leaving only a few young men glaring by the shrubbery, proving their manhood. One of them was the pisser.

"What happened here, marshals?" A sergeant was at Connell's window. We gave him a brief narrative.

"It was a damn stupid thing to do, coming in here to make an arrest by yourselves, parking up a deadend. This development is full of militants. We always use back-up units."

"It's finished now," I said. "We'll get the kids later. But I'd like to arrest that one buck over there, in the blue windbreaker. He pissed all over our car."

"Marshal," said the sergeant, "would you please mind getting out of here and leaving well enough alone."

Back at headquarters we had subpoenas issued for the parents of the four boys to appear in our office. They came readily and agreed on a personal recognizance arrangement to appear in family court with their sons. The kids ended up with five years' probation, which was not a fitting punishment.

For my next 10-13 I was with Ralph Graniela, a Puerto Rican marshal whom I admired more than anyone else for street sense. Graniela was so saturated in ghetto lore and habits that he had to be careful at formal functions that he did not go up to the mayor or someone and say, "Hey, do you be's the main man?" He became one of my closest friends. Patrolling the streets with Graniela was as safe as going with a platoon of Marines. Safer.

Only once did I think he was playing it too cool. We were cruising down Broadway in Brooklyn early on the night tour and just as we approached the intersection at Hart we saw a mob of people at the corner. They were milling around what looked to be four Puerto Rican men, faced off from each other. They were flashing long knives. It was not strictly our business but I had no complaints when Graniela bounced our car up onto the sidewalk, forcing a path through the crowd to the four men.

We bailed out, our guns drawn, shouting, "Drop the knives!"

Everyone was talking in a rush of Spanish and gesticulating at the knife-fighters, three of whom let their weapons clatter to the pavement. The fourth man turned and ran down Broadway, under the El.

"Grab him, John," Graniela shouted. "He's the main man."

I raced off after the guy, thinking that we were taking this too far. Why was I chasing this bastard down Broadway? I wondered angrily. We were supposed to be investigating a fire. Halfway down the block I remembered to pull out my badge and carry it in my left hand. There was a gun in my right and some cop might happen along and take me for the bad guy.

The fugitive was in poor condition. I caught him in two blocks and stuck my gun in his ear. If you had to pull a gun, I had learned, put it to effective use, short of squeezing the trigger.

Back at the car Graniela had the three other men draped over the hood, and I did likewise with the runaway. The crowd had reassembled, bigger than ever, and they were screaming fit to kill. One of the men over the hood was howling from a slashed arm and hole in his stomach.

"What's going on, Ralph?" I was yelling to make myself heard.

"I'll tell you later," he said.

"Later? For crissakes, there won't be a later."

I reached into the car and called a 10-13 on the radio horn. A second later, I swear, the sirens were whooping.

An unmarked car slammed up to the side of ours and disgorged a bunch of very seedy looking guys. I recognized them right off as members of the undercover anticrime unit at the local precinct. They were a beautiful sight, head-bands, long hair, and all.

"What's the scoop?" one of them asked in a slow drawl.

"Ask my partner," I said. "He speaks the language."

The cop talked to Graniela while I helped load the four men in the squad cars. When we got back into our vehicle and headed off up Broadway again, I noticed Graniela was grinning.

"Tell me, Ralphie, how come I'm suddenly in the middle of a mob that wants to eat my ass?" I asked him. "And who am I chasing? The good guy or the bad guy?"

"It's sweet, man, no hassle. It was a private fight—two against two. The guy you chased had been cut up three weeks ago by two of the others in a fight over a girl, so he brought along a buddy to even the score tonight."

"That's why we're nearly getting killed and calling a 10-13?"

"Yeah, now why'd you do that? The crowd was upset with the knife guys 'cause they were flashing their blades with kids around watching. That's why we intervened. You can't have men cutting themselves up in public. It ain't moral."

Graniela reached into an inside pocket and pulled out a knife, hefting it in his hands. He gave it a quick flick and a nine-inch blade snapped out and locked into place. On the handle I saw the panther emblem; the most prized knife on the street.

"Sonofabitch. You even took time to get a trophy. You and your damn trophies."

"It's the least they could do for our services," said Graniela.

I gave him a sidelong glance and got some taunt in my voice. "Typical Puerto Rican, got to have a knife."

Graniela looked over at me with real concern in his eyes. "Seriously, John. I was worried about you back there. I thought you were going to shit yourself in front of all those people."

By the time we reached the parkway we were both laughing like fools.

At 2 A.M. on a stifling hot August Saturday, in the ground floor apartment of an ancient tenement in East

New York, Zacharias Smith was arguing with his common-law wife, Elizabeth Davis. She had paid the rent last month, she howled; he had to do it this month. No way, he said, he fronted up for all the food. Okay, you lazy sonofabitch, she said, you get outta my home. The neighbors heard it go back and forth until Smith blew his fuse.

"You nagging bitch," he shouted. "I'm gonna cut you up!"

"Put the knife down," she shrieked. "You gone bananas again?"

"You'se gonna bleed, Liza."

Then came the sound of furniture crashing, doors slamming, and Elizabeth Davis was running down the tenement steps, yelling back at him, "I'm gonna get the man and he's gonna put you away forever."

Smith, a seven-time loser, convicted knife killer, a regular at Attica, did not chase her. He pulled out a zippo lighter and touched the flame to the sheets on the bed in the front parlor, repeated the arson in the two bedrooms, and set fire to the kitchen drapes.

The apartment was blazing and a woman across the street had called the alarm before Smith walked out into the hall. Upstairs residents tumbled down the stairs and ran by him to the street. He ignored them, shoulders hunched forward, a long switchblade knife gripped in his right hand. The glass door to the stoop was swung back in his face by one of the fleeing tenants. With a bellow of rage Smith drove his left fist through the glass, tearing lumps of flesh from his forearm.

A fire engine screeched to a halt outside as Smith hit the sidewalk. Firemen saw him, but the fire was getting away and they bent to their hoses.

"You'se done burned my house down," hollered the Davis woman. She had paused at the corner. A knot of people gathered about her.

"And now I'm gonna kill you," said Smith, advancing on her like a rogue lion, his left arm hanging loose and pouring blood.

An aide, who had been talking on the radio in the fire chief's car, saw Smith and heard the threat. "Brooklyn dispatcher. Get the marshals. We got the fire-setter here and he's got a knife." The aide, a fireman in his late thirties, scrambled out of the car and ran towards Smith, reaching for his shoulder.

"Hey, hold on there," he called. "Put the knife away."

Snarling, Smith turned back and lunged with the knife, opening a wide slit in the fireman's rubber coat. The fireman stepped back, watching, unarmed, afraid to try again.

By a stroke of coincidence we were traveling east on Atlantic Avenue, cresting Brownsville, when we heard the dispatcher's summons. Brian Rooney, a new partner, was at the wheel, and I was directing. This was my old stamping ground.

We hit Junius Street, two houses from the site of the fire, right at the corner where Smith stood glowering before a group of frightened people, his knife extended and gleaming in the streetlight.

"Holy, holy," said Rooney. "Guy's got a small sword."

He cut the motor and as soon as we got out with guns drawn we could hear Smith cursing at the top of his voice.

"You fuckers, I'll fix all of you. Come near me and I'll cut you'se fuckers up."

We came to the edge of the crowd.

"Who's the occupant of the apartment on fire?" I whispered to an elderly man who stood with staring eyes and a knuckle in his mouth.

"Coming this way. Lady in the red pants."

I picked her out backing off from Smith toward us and pushed through to her side.

"Hey, lady, I'm a fire marshal. I want to talk to you for a minute. What's your name?"

"I be's Elizabeth Davis and that there crazy man be's Zacharias Smith and he set my house on fire and I want him arrested."

"Okay, Elizabeth. If we arrest him you're coming to court to testify, right?"

"You got it. I want him buried."

I threaded through the mob, Rooney close by, to face Smith. I held my gun firmly, but pointed it down so as not to catch the light. Smith was no country boy. He would know that a short .38 was only effective at point-blank range, and even then you had to hit a vital target dead-on to drop a man. His bloodshot eyes riveted on me and I saw him come forward on the balls of his feet. He could beat me with the knife if he was quick. I had to mix him up.

"Mr. Smith," I said, "what's the problem here? There's no need for a knife."

"Come near me, honky motherfucker, and I'll cut your ass."

"Hey, c'mon, Zacharias, what kind of talk is that?"

"Don't call me Zacharias. I be's no friend of yours."

"Hey, Zacharias, what happened to your arm? You're bleeding like a sonofabitch."

"That be's my business. The bitch cut me. I'se getting even with her."

Rooney spoke up then. He had a strong, emphatic voice despite his pale baby-face and two hundred pounds of round flesh.

"Zacharias, you won't be able to stand up much longer if you keep losing blood. Let's get you to a doctor. We're not out to hurt you."

Smith cut at the air with his knife. "One step and I'll open your honky guts."

"Okay," said Rooney. "So drop dead. Bleed to death. Who gives a fuck. I'm going to the Hamptons after this tour."

The statement was so absurd, so unexpected—the Hamptons, for crissakes—that Smith dropped back on his heels. I leaped forward one pace and brought my gun up, cocked, aimed straight at his head.

"Drop the knife, Smith, or I'm going to waste you right now."

The crowd gasped and one woman screamed. Smith's knife clattered to the sidewalk. We were onto him in an instant, pinioning his arms and bundling him into the back of the car.

"Cumberland Hospital, fast," I snapped.

In the hospital emergency room, as a physician cleaned and sutured his left arm, I asked Smith, "Why did you try to knife the fireman?"

"I'se going away for good this time," he said wearily. "What's the difference. I cut somebody up. I'se so tired of peoples bothering me, pushing me around. I'se thirty-two years old. Twelve years I been in prison. What's it matter?"

Smith pleaded guilty to all counts and remains in prison to this day.

Ambush by militant black groups became a growing menace to fire marshals in the late 'sixties. We were essentially night-prowlers and we mostly operated in the ghetto streets. Nobody was more exposed. Extremists learned our habits, knew we came in on every suspicious fire—two men alone, carrying guns. Our guns and the money in our pockets was enough motive to be set up as a hit.

Brian Rooney and I walked into an ambush at 3 A.M. one fall morning in a rundown tenement in Brownsville. We had been asked to investigate a fire on the roof of the building earlier in the night. Three cases were already backed up and we wanted to delay the assignment until next day. But when a second fire was reported at midnight on the roof of the same tenement, we knew we had to give it priority.

The place was in darkness except for a weak light in the front hall that showed our way up the wooden steps and through the double doors, which opened at a touch of the handle. Just inside, to the left of the door, two men were sitting on the stairs, swigging from a pint bottle of wine.

"Hi, fire marshals checking the building," I said. The hallway was deserted, two closed apartment doors down the right side. "What's the problem in here?"

"I tell you, man, this place got troubles," said the taller of the wine drinkers, getting to his feet. "Some dudes they'se trying to burn the roof off'n this good old house. Two times tonight the fire boys is coming. Me 'n Andy here so nervous we cain't sleep."

"That's what we're here for," I said. "Want to find out who's setting these fires. You see any strangers hanging around tonight?"

"Yeah, man, we can help you'se on that." Andy was on his feet and talking directly at Rooney, who was on my right. Andy had the bottle. He took a pull and thrust it at Rooney, who was obliged to step back a few paces.

"Wanna drink, marshal?"

While Rooney was shaking his head, Andy, a guy in his mid-twenties, close-cropped Afro hair, shot a furtive glance at his pal. "I dunno," he said. "Man could get burned opening his mouth hereabouts."

He put his free hand lightly on Rooney's shoulder and motioned him down the hallway, as if out of earshot. His voice became a murmur and Rooney bent to listen.

"He's just jiving," said the tall man. I forced myself not to turn away from his sour breath. "That Andy's always too juiced to see nothing. You a marshal, you'se know why them dudes is burning down the houses. This here's the roof job paid by them sonofabitch landlords. Decent folks ain't safe no more."

"And we want to make it safe," I said, "but we gotta have information." Out of the corner of my eye I saw Rooney going further and further along the hallway. My right hand was in the big flap side pocket of my new chocolate-brown leather jacket. I had moved my gun there before entering the building. My hand tightened around the butt.

The man in front of me kept jabbering on about crooked landlords, talking sing-song, his head bobbing.

And then it happened. The first apartment door opened

and the second an instant later. Two men appeared sleepily in the door of each.

"Hey man, what's going on here?"

The trap was set perfectly. Rooney and I had six men between us. At a signal they would rush me or him, or maybe both. If either one of us was fast enough to pull a gun, they would flatten on the floor. Rooney and I would be shooting at each other. They knew the routine was to empty the gun at the first fusillade.

The leader reached over and took the lapel of my jacket between his fingers. "Man, that's one beautiful coat. That be's real leather."

I held my eyes on his. We were already on borrowed time. I raised the barrel of my gun no more than an inch inside my pocket, pointing it directly at the man's belly. The movement was lost on the others in that dim light, but my man saw it.

"You be's ready?" It was scarcely a whisper.

"You bet your ass I be's ready," I said, also softly.

I took two short, rapid paces backwards, my hand still on the gun in my pocket. At that range the .38 could deliver five slugs within the space of a half dollar in the man's heart.

"Everybody on the floor," I shouted. "Brian, get over here. Let's go."

Rooney looked up in surprise. Andy cursed.

"You crazy, man. We just here rapping."

"Down!" ordered the leader, spreading himself on the floor. "The man *knows*, goddam it, he *knows*."

Out in the car I put my hand over my heart to stop it from jumping out of my breast.

"Never do that again, Brian," I said angrily. "As long as you work with me or anyone else in this goddam jungle, never allow yourself to get separated."

"I was watching it like a hawk, John. Everything was under control."

As Rooney moved our car out from the curb a police patrol car, the roof lights dark, pulled up alongside with a squeal of tires. A uniformed cop rolled down his window and stuck his head out, glaring at us.

"You fire marshals are gonna get wasted one day," he barked. "Three o'clock in the morning and you're walking into these joints as if it was Bloomingdale's at high noon. That very building you came out of, night before last, we had a cop gunned down on the third floor. Not dead, thank God."

"We'll be more careful, officer," said Rooney. I could see his face go into a grin. "No trouble tonight."

Rooney and I finished that night tour with our feet dragging and I don't think we were champing for too much action when we reported in after a two-day break to work a Manhattan watch. Manhattan is nowhere near as busy for marshals as Brooklyn or the Bronx and we happily sat around headquarters drinking coffee, smoking cigarettes, and making like we were in deep conference. It was raining outside, a drowsy late spring rain with drops the size of marshmallows.

The rain stopped at two o'clock and I called Flo and suggested a barbecue dinner for later and a bottle of good Italian red. At two-oh-five we got an alert to check a positive incendiary fire at a third-floor apartment on Mulberry Street.

We were there in five minutes; a really nice apartment house, biscuit-colored brick, eight stories, a doorman, a tiny fountain in the lobby, even two elevators. The gray, metal-sheathed door to 3C was scorched black and the carpet in front singed to the floor timbers. To one side of the door, still soaked by the water from the fire extinguishers, were crisps of cardboard, a mangled umbrella, and a woman's molten rubbers. Judging by the severity of the burn marks in the floor underneath, we figured someone had poured gasoline into a box of rain gear and set it alight. The alarm must have been called immediately because the fire had been confined to a small area. Seeing the umbrella and rubbers outside, we guessed the occupant had been at home, and smart enough not to open the door and give the fire draft.

A pretty young woman in a plaid skirt and white blouse answered the buzzer. She had soft red hair and her green eyes were wide set under broad, rounded cheekbones. There was a smudge of freckles in her fair skin.

Catherine Mulvaney led us into an immaculate green-walled living room, a rich tan rug underfoot and low modern furniture at the edges, and introduced us to a short, dark man named Alfred Marino. He nodded and looked surly. The girl sat down on a seed cotton couch and reached for a cigarette from an open packet on the glass coffee table.

"The man you should be looking for is Tony Abborelli," she said. "He's the one set my door on fire. He's a real bad ass."

Rooney's eyes flickered at the expression. It sounded coarse in this room, coming from this girl.

"We had been going together but I broke it off because every time he got booze in him, which was often, he'd start pushing me around. I had enough of that from my own father."

Marino came into the center of the room holding his hand up. "Knock it off, Cathy, these guys don't want your life story. Jesus, one lousy little fire and you're making a big deal."

The girl ignored him. "Tony came busting in here this afternoon telling me he was gonna break both my arms—"

"Okay, okay," interrupted Marino. "Enough of the crap. Just tell 'em you smelled smoke and called the fire department."

The Irish girl drew on her cigarette, trying to get her line of thought going again.

"Tony went away again and then ten minutes later he's thumping at the door. I told him to get lost, I wasn't opening the door. He says if I don't open up he'll burn the whole building down."

"You're running off at the mouth, Cathy." Marino's voice had menace in it this time.

"Hey, pal," I said to him. "Who the hell are you?"

"I'm a friend."

"He's a friend of Tony's," said Catherine Mulvaney.

"You're not related," I said, "so do us a favor and leave the apartment."

"And let you snoops get some cock-and-bull story from a crazy broad like this one here? No deal."

Rooney's face flushed with anger. In one bound he was onto Marino, spinning him around, one huge fist bunched on his collar and the other seizing the man's belt. Marino's feet left the ground as Rooney marched him through the archway and out into the kitchen. The window stood wide open.

"Out you go," cried Rooney. And with that he released the man's collar and flung him through the window. Martino dangled three floors above Mulberry Street, held only by Rooney's grip on his belt.

"No, Brian, no!" I charged into the kitchen and dug my fingers into the cloth of Marino's jacket, yanking him back inside.

"Pal," I said, breathing heavily. "Don't you think you'd better leave?"

Marino was off and running for the front door as if pursued by the hounds of hell.

"Sorry about that, miss," said Rooney pleasantly, returning to the living room. "May we have more of the story now?"

The girl seemed to take the incident in her stride. She must have kept some rough company in the past. She had not even moved from the couch. At our prompting she continued the story, saying that ten minutes after she did not open the door to Tony Abborelli she saw flames licking under the door and phoned in the alarm.

Rooney left us then and checked tenants down the hall. A woman in 3D confirmed that a man had been banging on the door of 3C and threatening to burn the building. The man in 3F said he had seen a man splashing the contents of a silver can into a box by the Mulvaney apartment. He had been peeping through a crack in the door and closed it abruptly at this point. It was none of his business, he said, and he did not want to get involved.

"Involved!" Rooney was a man with a short fuse. "You're lucky you weren't roasted alive."

My partner was back in 3C with his report when the telephone rang. It was on the rug by the couch. Catherine Mulvaney reached for it, listened a moment, and started jabbing her finger at me and then the phone.

"Abborelli?" I mouthed the words.

She nodded at me and said calmly into the mouthpiece, "You're sick, Tony. You oughta have yourself psychoed."

"Get him to talk about the fire," I whispered. She held the receiver out from her ear so that I could hear, too.

"Why did you set the fire, Tony?" she asked.

"You bitch." The answering voice was agitated. "I should have burned you out into the street. I catch you on the street I'm gonna kill you."

Then the line went dead.

"Can you use that?" asked the girl, dropping the receiver back into the cradle. She was twenty-three, at most, this girl, but she had walked some tough streets.

"Yeah, sure," I said. "I got your consent to listen, so what he said is admissible evidence in court."

She told us Abborelli lived with his parents in a house in Flushing, Queens, and to be careful because he carried a gun. She also gave us a snapshot of him. I called Flo and canceled the barbecue, then we drove on out to Flushing, taking the 59th Street bridge and hitching up to Northern Boulevard.

It was a Tudor-style home, a big willow in front that was already in full leaf. The mother opened the front door and we identified ourselves and said we were investigating a fire about which her son Tony might or might not have information.

"He's not home," she said. "and I don't know where he is. You'll have to look up one of his so-called friends. He must be staying with one of them." She paused and looked at me with worried eyes. Perhaps she saw in me and my Italian cut a person to confide her troubles in. "Tony's been a problem ever since my husband and I separated. He was a no-good man and I fear his son is going to turn out the same way."

I gave her a sympathetic look and handed her a call slip. She promised to telephone if her son was in touch with her.

The Mulvaney girl had told us that Abborelli was working for a taxicab company, also in Flushing, and we went there next. The manager was in a glass office up a short flight of stairs from the main floor of what proved to be a long, double-fronted garage. A painted sign at the entrance said SAL'S LIMOUSINES. Three unmarked sky-blue Chevrolets were parked inside the open swing doors and four or five mechanics worked on other cars in the rear. It was the sort of place that made you wonder what else the owners did for a living.

"Abborelli?" said the manager, a hairy man with brown teeth. "Sure, he worked here. A punk. We fired him. He lived with his parents. Other than that I know zilch."

We drove to a telephone booth on Parsons Boulevard and called Catherine Mulvaney for some addresses and information about where Abborelli took aboard the liquor for his mean drunks.

"I've been trying to reach you," she said. "An hour after you left I saw Tony standing by a car outside my apartment house. He got in after a while and I swear I saw that car pass by about five times. It was a light blue Chevy."

"Sky-blue?"

"Yes. I also got the license number."

I wrote down the number and also Alfred Marino's address and the names of three bars. My notepad had the number for Sal's Limousines and when I hung up from the girl I dialed it.

"Why didn't you tell me Abborelli's still driving one of your cars?" I asked the manager.

"Because he's not."

I read out the license number Catherine had given me. "One of yours?"

"Shit, yes. We reported it stolen two days ago. Abborelli did that?" He was incredulous. "Punk's got some helluva nerve. They'll take his head off."

"Who are they?"

The manager let the question hang there, too long. "The cops," he said limply.

"What do you think, Brian?" I said when I got back in the car. "It's seven and we're off duty but I got a few leads here and there's no telling what this guy Abborelli might do to the girl."

"Let's stay with it," said Rooney.

We drove the patrol car back to headquarters, switched to Rooney's red Camaro, and returned to Queens.

Marino roomed in a comfortable old residential hotel on Metropolitan Avenue, but his room was vacant and when he did not show in two hours we started on the bars. The bartenders at the first two acknowledged that Abborelli came in often and we wasted hours nursing a drink at each, waiting and questioning the other regulars. No leads. It was well after midnight when we reached the third place, The Velvet Swing, on Woodhaven. The bartender got very fidgety when we flashed our badges and asked if Abborelli had been in.

"Abborelli. Abborelli." He repeated the name as if it rang the most distant of bells.

I shoved the snapshot we got from the girl in front of him. "Oh," he said. "That Abborelli." He was the worst actor I had ever met.

"Pal, you got information, you spill it out or you're gonna get your ass locked up," I said.

The place was empty except for two men sitting apart near the door and well into their cups. The bartender moved to the far end of the counter.

"Tony left here an hour ago with the owner. He's staying over at his house." We had trouble hearing him his voice was so low. "If you tell my boss that I told you that, I'll get broken in half."

He gave us the owner's name, Matthew Green, and an address in Ozone Park.

"Let's kill a couple of hours," I told Rooney outside. "If we hit the place now we'll have to make enough noise to wake up the whole house to get a door opened. People are in their deepest sleep at this hour. Towards five they're close to waking and we'll be able to alert someone without too much racket."

"Agreed," said Rooney. "I get the impression that Abborelli's playing at the fringes of the mob. In fact, I think he stole one of their cars, which makes him the prize hothead of all time. He's gonna be as jumpy as hell. He carries a gun and I'll wager he's about ready to be shooting at shadows. We'll never take him if he has any warning."

We crossed back to Manhattan for bacon and eggs at the Market Diner, drank four cups of coffee, and cruised slowly to Ozone Park, searching the streets there for the heisted sky-blue Chevy. It was nowhere in sight. We parked down the street from the bar owner's address and picked his place as either the first- or second-floor apartment of a two-family garden apartment that stood at the edge of a dozen similar units.

At five we walked over to the front door and gave three quick jabs at the button under the black name strip that said Matthew Green. He lived on the first floor. Within a minute the door was opened by a husky man in a teeshirt and boxer shorts.

"You Green?"

"Yeah. What is this?" He saw the gun in my hand and reared back.

"Where's Tony?"

"Ain't nobody here," said Green.

"You want to get blown up, you tell us there ain't nobody here." Green looked at Rooney, then back to me. I was wearing white belt, white shoes, red pants and a dark blue jacket. The stubble showed on my chin. I did not look anything like a cop.

"God, no!" Green's voice was strangling in his throat. His eyes were popping. "Back room. Tony's in the back room. I don't know nothing about anything."

We pushed him aside and walked through the front room, down a hallway to a bedroom beyond. Twin beds; one empty, a man asleep in the other.

Rooney stood at the front of the bed, his gun aimed at the mat of black hair on the sleeping man's chest. I walked to the head and eased the barrel of my pistol into the man's ear. His eyes opened, but he did not move.

"That's it, Tony, don't move an eyebrow."

I reached my arm under his shoulders, holding his arms to his side, and scooped him out of the sheets. My hand went under the pillow, then into the space between mattress and boxspring. I felt metal and pulled out a long-barrel .38, which I dug into my belt.

"What's going on?" Abborelli was on his feet, naked, his eyes on Rooney now as the big Irishman held the gun on him with both hands.

"You know why we're here," I said. "Get dressed, Tony." His clothes were neatly laid out on the other bed. I handed them to him, item by item, running my hands through the pockets. He dressed without speaking, watching Rooney's gun. He was a handsome young guy, with thick black hair falling across his forehead.

I retrieved the car keys from the side pocket in his jacket before I handed it over.

"Where's the car, Tony?"

"What car?" It was a weak stall.

I bounced the keys in my palm. "Take us to the car, Tony."

He went meekly out of the house without saying a word to his host. Green was in the front room sitting in a corner chair, still in his underclothes, motionless as a statue.

Abborelli sat with me in the back seat of Rooney's car and guided us through back streets to a supermarket lot where he had left the stolen Chevy. I could see his body trembling and every time he gave directions his voice quavered. I was feeling sorry for him.

"You handle the fire report on this one, Brian," I said when we saw the supermarket ahead. It was time to rescue Abborelli from his terror.

"Sure, John. I took full notes from Catherine Mulvaney."

Abborelli stiffened at my side. "You guys are cops!"

"Fire marshals, Tony."

"Sweet Christ. You're taking me to jail. I thought you were from the mob. I thought I was a goner."

"Sorry, pal," I said, "but we wanted the larceny as well as the arson in Mulberry Street and there was no way you'd tell a couple of marshals, or cops, where the car was. Without the car we couldn't have made the larceny charge. See, Tony, we got this thing about people who set fires. We want 'em off the street, and in a case like yours, the longer the better."

Abborelli was only half listening. He was smiles from ear to ear. "Hey, give me a cigarette. What a relief, man. Arson, for crissakes, and larceny. That's nothing like being wasted."

"Yes, it's something," I said. "You're gonna go away for quite a spell."

Abborelli was laughing then, and Rooney and I could not help laughing along with him. At dawn that morning, in a precinct in Queens, we booked the happiest man in the city.

In the acting department we found our match in Jesus Umbrerra, a diminutive Puerto Rican guy with an enormous Jerry Colonna mustache and a smile as wide as a toothpaste commercial.

I was working with Rooney again, our spirits in tune with a brilliant spring night, when we received a call to an apartment house on Bedford Avenue in Greenpoint.

A fire captain was waiting impatiently outside the

building. "Come on you guys, I gotta go. We're holding a suspect in the cab of the engine. The neighbors were beating him up for setting the fire and we took him in for safety. It was only a mattress fire at that."

I opened up the cab door and saw a little figure huddling on the floor, his face cocked around and beaming at me.

"Hi, pal, what's your name?"

"I am Jesus Umbrerra," he announced.

"Okay, Jesus, out of the cab and into our car. We'll look after you."

Rooney took him into the Plymouth and I went off to make an inspection. The captain showed me a burned mattress lying on the curb where the firemen had dragged it from a second floor apartment. I walked into the old brownstone and was escorted to the scene of the fire by the landlady, a shapeless woman who called Umbrerra a little rat. She herself had let him into the place, which was rented by his daughter, because nobody was home. Next thing, she said, he was trying to burn the building down.

On the wall of the large bedroom off a spotlessly kept front parlor I noted the funnel-shaped scorch mark that would have come from the now unmattressed bed. There was no other damage. A can of Miller's High Life was on the floor at the side of the bed. I grinned at that. A million Puerto Ricans in New York drink Miller's. It's a status symbol. The ring cap on this can was gone and it felt about one quarter full. I stayed there a few minutes, letting the landlady's rasping voice bounce off my ears, while I thought it out. If Umbrerra had come to burn the apartment deliberately, why did he pause to drink beer?

We took our suspect to the squad room at the precinct in Greenpoint. The room seemed even more unloved than the one at the 73. The harsh glare from 100-watt bulbs screwed into porcelain wall sockets reflected off the pale green paint and threw a ghostly pallor over our faces.

"What do your pals call you, Jesus?" I asked.

"Everyones call me Herbie," he said. It was difficult to watch his eyes and not the handlebar mustache flopping up and down under his nose. "You'se call me Herbie."

"What happened, Herbie?"

"Mans, I do nothing wrong. The peoples punch me when I come outa the building, but I do nothing wrong. See, I go to my common-law daughter's apartment. It's locked so the landlady lets me in."

"Hold it" said Rooney. "What's a common-law daughter?"

"She's my common-law wife's daughter." Umbrerra made the explanation, then flashed his magnificent smile.

"We're getting into some beauty here, John," said Rooney. "I can see it coming."

"Okay," continued Umbrerra. "I go in and sit down a little while. Nothing to do. I opened the refrigerator and got a can of beer and sat on the bed drinking it. I lit a cigarette and I lay back and must have fallen asleep. I wake up and the mattress is on fire. So I run downstairs to get the landlady and she starts punching me and telling me I made the fire on purpose."

"That, Herbie, is a bullshit story," I said. "What's the game? Did you have an argument with your common-law daughter? Or maybe it was your common-law wife?"

"Daughter, no," he said. "My wife, I am always arguing with her. But that's not saying I lit any fires deliberately. I would never make fire in a house. Never. I swear it on my wife."

"Herbie, you don't like your wife," said Rooney.

"All right, then, I swear it on my daughter."

"She's only a common-law daughter." Rooney was getting a charge out of this interrogation.

Umbrerra gave another of his triumphant smiles. "I swear it on your daughter."

"No, you won't, Herbie," said Rooney. "I'll break your head for swearing on my daughter."

"I mean nothing bad," Umbrerra protested. "I swear it on the saints, on the Madonna." His eyes shot around the room, and arrested on the light in the corner. "I swear it on that lightbulb."

We broke up on that one, and Umbrerra beamed like the sun at our enjoyment. He had us. The lightbulb and the evidence of the beer can convinced us that Umbrerra was telling the truth. And even if he was not we had nothing to book him, nor the remotest chance of getting anything.

"Come on, Herbie," I said. "You're coming with us."

We drove him down to the Brooklyn-Queens Expressway, to the Gowanus, and then all the way around the horn to the Shore Parkway, stopping in a deserted area by Floyd Bennett Field.

"This is where you get off, Herbie," I said. "If you go near Bedford Avenue again tonight the neighbors will kill you, and it won't look too good for us either. You got any money?"

"Nothing," he said. "Not a penny. That's why I went to see my common-law daughter."

I counted two dollars out of my wallet, far short of a cab fare back to Greenpoint, and gave it to him. "This is so you're not stuck."

"Thank you mans," he said getting out of the car. "I won't go back to that apartment. I swear it. I swear it on the sky, on the stars . . ."

We drove off then, leaving the strange small person of Jesus Umbrerra standing at the roadside, his white teeth flashing and his mustache bobbing as he made his vows to the world that he was an innocent of any foul intentions.

"It's spring," said Rooney. "I swear it."

Chapter 8

The Case of the Mail-Order Bride

IT is a cliché that true detective work is ninety-five percent a dreary routine of ringing doorbells, asking questions, and assembling facts piece by careful piece. The master at this in our division was Phil Winters, and I credit him for showing me the importance of always going the last mile on a case.

Winters, prematurely bald, as fat as Jackie Gleason, was known to everyone as "the chief," a name that went back to his firehouse days. They said you could have a five-alarmer and a clutch of legitimate battalion chiefs present, but Winters still tried to call the shots. He was not disliked for it. The guy just happened to have an all-encompassing, wide-angle mind.

Late one mild Saturday night, I was patrolling with Winters in Car 50, which was doubling up on both Brooklyn and Queens. We had a standing instruction to respond to any fire that went to a second alarm, and this night we received an urgent radio call to drop everything else and investigate a second-alarmer in Sunnyside, Queens. The Queens dispatcher had not finished giving us the address before Winters had swung the car in a tight U-turn on

Eastern Parkway, in Brooklyn, and was belting off down the eastbound road.

I was the radio man. "Car 50 to Queens dispatcher. Acknowledge second alarm Fourth and Trinity. Responding. Man, are we responding!"

"Ten-four. Confirming one 1045 code one, approximately twelve code twos and threes."

Winters made a sound that was halfway between a groan and a sigh. Code one was a death, the other codes meant injuries serious to critical.

Winters' left hand flicked to the siren switch, then back to the wheel. I put the bubble light on the roof. The Plymouth lifted on the thick police tires and we were quickly to eighty miles an hour. Winters kept his back straight and supple in the seat. He was controlling the big car with his wrists, rifling it down the center lane, the streetside maples a blur in the darkness and the high mercury lights of the parkway darting overhead. Up ahead, warned by our siren, the crimson tail glows of other cars scattered to the right and left.

The radio crackled again. Fourth and Trinity had gone to a third alarm.

"Give them an ETA, John," said Winters. "Fifteen minutes."

I radioed in, trying not to sound too agitated. Winters had poured on more speed. My heart was banging like a drum. I squirmed closer to the passenger door for fear he would hear it, or maybe see my jacket jumping in and out. Some Winters. His eyes were narrowed, focused at the far end of our headlight beam, but his chubby face was impassive and his plaid jacket lay flat on his chest. He was not the man to ask about a cure for a galloping heart, my single major handicap as an intrepid fire marshal.

Winters cut up Bushwick, bounded onto the Brooklyn-Queens Expressway, and had us into Sunnyside within the fifteen minutes. It was 2:55 in the morning. The brownstones and garden apartments were mostly dark as we scooted through the outskirts. By the row of brown-battened stucco stores on First we caught up with an ambulance and two fire engines. A police car flashed in from the side alley and soon we were part of a noisy rescue squadron summoned from all corners of east Queens. House lights were on now along both sides of these streets. I watched the faces at the windows, feeling superior to them. It's all right, I wanted to tell them, Winters and Barracato are on the way.

"Jesus!" Winters had wheeled into Trinity. The street was pandemonium. Fire rigs lined the roadway for a hundred yards, red roof lights spinning, headlights blazing into the brick residential buildings on either side and glaring off the wet pavement. Hundreds of people with white, night faces were walking among the engines, stepping over the fire hoses. A weary voice on a bullhorn somewhere up the block was saying, "Everyone back, pulease, everybody back."

The fire rigs had left access for the ambulances, which were still behind us, and Winters pulled off onto the sidewalk. We went forward at a trot to the advance engines, parked in an arc around a three-story apartment house. The building hissed with smoke and steam, and tongues of flame spurted from windows, but I could see that the main force of the blaze had been knocked down. The inside of the building seemed almost completely gutted.

On the roadway in front of the apartment house, firemen had carefully laid out the still forms of men, women, and children who had been overcome by smoke. Some had leaped from windows, others had been carried out. The lifeless body of one woman, burned pink and black, was being folded into a gray canvas body bag. Firemen knelt over other victims, mouth-to-mouth, breathing life into their lungs.

A cry from the crowd of spectators on the opposite sidewalk made me look back to the building. Through the gaping entranceway I saw a figure on top of the hall stairs; a fireman cradling a child. He must have gone in through the roof, descended to the second floor for the child, and found his retreat blocked by flame. He had made his way down. Now he was trapped. Fire was curling around the steps from the few timbers that remained of the base structure. The staircase would never stand his weight.

A ladder company man by my side, his eyes red-rimmed in a soot-blackened face, took a few steps forward, then stopped. "Lennie!" The word was a sob in his throat.

The fireman on the stairs hugged the child to his body and covered it with a flap of his rubber coat. He started down, hunching his shoulder against the flames. Two steps, three step. Away down the street I heard the wail of ambulance sirens, the shouts of men. Four steps. The fireman put one heavy boot on the fifth step and had the other lifted when the whole stairway collapsed under him in a great cascade of fire and sparks. For a moment he was lost to sight in the falling, burning timbers. Then we saw him, erect and holding the child, stamping out of the fire and into the street.

A cheer went up from the crowd and I felt my eyes flood with tears. I spun around to summon up some professional detachment. My eyes fixed on a small, dark-haired man who was weaving his way past the front clutter of fire engines. His face was bloodless with grief. With lunging steps he went toward a row of five children set down on the pavement next to the dead woman in the canvas bag. It had been corded up with white nylon rope like a freight package.

"Heidi?" He was a few paces from the first child, a little girl no older than six, raised on one elbow and looking about her dizzily.

"Heidi." The man was crying, cupping his hands around the little girl's face. He looked to the next child, then the next, calling their names. They were recovering from the smoke inhalation and stirring under their gray blankets.

"Papa!"

My mouth went as dry as ashes. The man must have this minute returned home. These were his children.

I looked to the canvas sack, then back to the man. The

pain in my breastbone made me suck in a deep breath. The man walked slowly along the line of his children, seeing the bag, brushing past the restraining hands of the firemen who had been tending the little figures on the roadway.

"Is Momma?" The man had reached the oldest boy, a son in his early teens. The father crooked an arm toward the body bag, spreading his fingers. His voice rose to a shout. "Is Momma?" The boy, who had been sitting with his head between his knees, struggling for breath, turned his face to his father and in a soft voice he said, "Yes."

Two firemen, massive beside the father in their heavy, wet turnout gear, supported him as he fell to his knees whimpering. A strange, small sound escaped from my mouth, and I turned away, not bearing to watch any longer. I walked toward a square of grass at the front of the apartment house adjoining the devastated building. The fire was conquered, but it would be a few minutes before the smoke would clear sufficiently for us to go inside.

Over on the grass a fireman was crouched by a black bundle, and when I got closer I saw that it was a boy of about eleven. His hair was singed to small tufts and his scalp and face were raw pink. The boy had one scorched arm raised limply.

"Water," he whispered. "Please, a drink of water."

I knelt beside the fireman.

"Where in God's name are the ambulances?" he asked.

"They're here," I said. "You'll have the stretchers in a moment or two."

The dying boy on the grass moved his arm toward my voice.

"A drink of water, please mister."

"Soon," I said. "We're just waiting for the ambulance men."

The boy was my own son's age, only John was home in his bed on Staten Island with his head on a pillow and a new football helmet perched on his reading table.

"It's gonna be okay, son," I said. Over my shoulder I saw the white forms of the ambulance men gathering up the kids on the roadway.

"Over here!" The fireman bellowed at the nearest crew, and two stretcher-bearers changed direction and came over to the grass courtyard. Behind them, a man with a bald head was glancing about impatiently. I scrambled up and walked over to him in a hurry.

"Ready to go, Phil."

He gave me a sharp look, appeared to begin a comment of some sort, then shrugged.

Winters led the way up the littered front steps and into the hallway of the smoldering building. Hoses were still playing through the upper windows, and water washed down on us through the broken timbers of the second floor. Neither of us had stopped to put on rubber coats and helmets; a breach of regulations.

The instant we stepped inside we smelled gasoline, and we could see by the deep charring of the walls at the front of the hall that someone had splashed it around liberally. The wall tiles had dropped off in the heat and the plaster behind them had been consumed by the fire. Without an accelerant the blaze would have given the walls a light scorching and skipped on in search of more combustible material. I felt sick to my stomach. Thirty people, I estimated, had been asleep in this building when some mindless savage had deliberately set this fire. One woman was dead, a little boy was dying in agony, and at least fifteen others would carry physical and mental disfigurements from this fire for the rest of their lives.

The burning seemed fiercest by the door of the first apartment on the left as we came in, so I rummaged around in the debris on the floor here while Winters went further down the hall. A few feet from the door I found a coconut webbing mat, barely touched by the fire. It had been washed clear by the first of the high-pressure hoses brought into play. One sniff at the footmat and I walked outside to the nearest police patrol car.

"We want everybody out of the area, officer," I told the uniformed cop standing by the driver's door. "Get the front of the building roped off. This is now a declared crime area."

"What have you got, John?" Winters had joined me on the outside.

"Look at this mat, for crissakes. You could stuff it in a tank and drive around the city. The arsonist's target was apartment 1A."

Winters went off to inform the battalion chiefs that the fire was definitely set, and how and where it was done, while I walked back behind the police lines and asked among the crowd for the people who lived in 1A. Ironically, they would have suffered the least, for the full fury of the blaze had gone to the upper floors. A lad in pajamas and raincoat pointed me to a squat, black-haired woman who was wringing her hands and babbling away in a language I identified as Italian. She was aged somewhere between twenty-five and thirty-five, a lumpy woman with protruding eyes, wearing a faded blue velvet robe. A gaunt man with gray hair stood by her side with one arm loosely around her shoulders.

"Excuse me, ma'am, I understand you live in 1A." My normal speaking voice comes over as kindly, in a low, throaty way, and it shushed the woman for a moment. Then she saw my shield, pinned on to the breast pocket of my jacket, and her wailing resumed.

"*Joyelli anyelli,*" she cried. "*Mia joyelli.*"

The words, which I could only remember and write down later phonetically, meant nothing to me. I asked her to speak English.

"She no speak English," said the gaunt man. "I am papa. She crying for jewels, *joyelli,* jewels. Have they destroyed?"

"We'll look for them later," I said with annoyance. I had not been able to get out of my mind that dying boy on the grass with his scorched arm raised and asking for water. The woman was concerned only about her trinkets.

"Just you and your daughter live in 1A?" I asked.

"Yes."

"Well, I hate to say this but I think somebody's got a vendetta going for you. The fire was purposely set to burn you out. Have you got enemies?"

"No," he said.

"Is anybody mad at you?"

"My daughter's fiancé. She break off. He made very angry."

At this point Winters walked up with Vincent Canty, who was called in on all major fires, and we agreed to continue the questioning at the local precinct.

"This is one for you, John," said Canty. "You're the only one of us who speaks Italian."

"Right," I said, trying to sound complimented.

I plodded off after Winters, who was walking between the old man and his daughter. I had a problem. On my application for the fire marshal's job I had falsely stated that I spoke Italian. Nobody had ever caught me in the lie—not yet.

Canty gave me the chair opposite the woman in the precinct squad room and suggested I start the interrogation. Winters stood by with two homicide detectives.

I cleared my throat and said, "Casa . . ." It was one of the five or six words in my Italian vocabulary.

Fortunately, the woman broke in and began talking rapidly.

"Say," I said to the father in sudden inspiration. "Where are you folks from in Italy?"

"Milano," he said.

"There's the problem, chief," I said, wheeling to Canty. "My parents spoke a Sicilian dialect. She's never going to be able to understand me."

"John's got something there." Winters came to the rescue. "There's bound to be someone in the precinct who speaks northern Italian."

There was indeed, one of the drivers downstairs, and he heard a bizarre story from Sophia Tracelli, the daughter.

A young man named Guiseppe Frassa, who ran a small plumbing business in Sunnyside, had sent his uncle in Milan one thousand dollars to export him an Italian bride. For obscure reasons known only to the uncle, he chose Sophia, who held out for another thousand so that she could bring her father. Frassa paid up, and, when they duly arrived, installed them in the apartment on Trinity Street. He slipped a diamond ring on Sophia's fat finger and prepared for the wedding. In the opinion of those of us listening to the story as it was being interpreted, Frassa would have done better to escape the contract and run to California. Yet it was Sophia who got fussy. She found Frassa cheap, and not the rich American she had been led to believe awaited her caresses in the magic city of New York. She broke the engagement. When she refused to return the diamond ring, according to her, Frassa raged and shook his fist and stormed off into the night. She had not seen him for days, but understood he was staying with his sister in White Plains. The ring, of course, was the prime reason for her wailing at the fire. She had not spared a thought for the suffering around her.

While Winters returned to Trinity Street, I drove up to White Plains—to the address Sophia's father had given us—with the two homicide men. It was seven o'clock in the morning when we knocked on the door of a cedar-shingled split-level in a pleasant sidestreet. A sleepy woman in a man's robe answered our knocking, and, at our request, she called to her brother. Frassa appeared in the doorway in flannel pajamas—a plain man of average height and weight with a large nose, black-framed spectacles, and a dark stubble on his chin.

"It's very early, sir, and we apologize for the intrusion." I was extremely polite. "We're police officers and we believe you may be able to help us in a case we're working on. We don't want to disturb the household. Perhaps you would get dressed and come to the station with us."

"You got the wrong man," he said.

"We'll explain in the car," I persisted.

"Okay," he said irritably. "Give me a minute."

Frassa rode with us in silence until we turned down the Post Road, heading for New York City.

"Say, what is this?" he yelped. "Where you taking me?"

"Queens," said the cop in the front passenger seat. "Just sit back and enjoy."

Under questioning at the Sunnyside precinct, Frassa denied setting the fire, denied being anywhere near the apartment of his wretched former betrothed at the time of the fire. He was asleep, he said, at the house in White Plains.

His sister and her husband verified that he was staying there that night, though they had gone to bed at midnight and they conceded that he could have sneaked out some time after one o'clock in the morning and made a fast return trip to Sunnyside in his van. The van, a short-axle, late model Ford, pepper green, had been parked at the street end of the driveway. The sister's bedroom was in the back of the house.

On the following day, Monday, the newspapers gave the fire prominent space. Two people had died, the woman and the little boy, and twelve were injured. The man I had watched crying over his children, collapsing at the realization that his wife was dead, was identified as Dov Shaluskarim, who had brought his family from Albania less than four years earlier. He worked as a chef for a Woodside restaurant and when I saw him he was arriving home from work. The brave fireman with the child in his arms was Leonard Velox, in good condition at Queens General.

One of the newspapers carried a picture of Winters and me examining the rubble of the fire. We seemed a couple of professional stone faces, immune to the heartbreak. Okay, I thought, these two unfeeling ballbreakers are now going to do their chosen work.

We had motive and opportunity—barely—on Frassa, and by canvassing the hardware stores in White Plains we came up with the means. A clerk at the Windmill store identified Frassa from our Polaroid picture as the man who purchased two red, two-gallon jerry-cans on the previous Friday. A Mobil gas station attendant remembered filling up two such cans for a man who looked like Frassa and who was driving a green Ford van.

Frassa scorned this circumstantial evidence. His gas gauge was unreliable, he said, and he had decided to carry the gas in case he was stranded on the road.

"We're running out of chances," I complained to Winters.

"Bullshit," he said. "There is a routine to the life of a street even at two-thirty in the morning, when this fire was started. People watch late-late movies, people get out of bed to take a leak, insomniacs wander about their apartments smoking cigarettes, newspaper delivery men come by, men come home from bars, shift-men come home from work. We're gonna study the action on Trinity Street at two-thirty every morning until we find someone who saw something last Saturday—like, for instance, a green Ford van in front of number thirty-two."

Trinity was a north-south block, two hundred yards long. The brick, six-family apartment house that had been gutted was midblock on the east. Similar places flanked it right and left, and beyond these, and on the opposite side of the street, there were rows of crisply kept two-family wood-frame homes with broad front steps and verandahs. On the next three nights we staked the street and stopped everyone we saw from one o'clock on. Milk delivery vans, we discovered, still made rounds in Sunnyside in the wee hours of the morning. A *Daily News* delivery truck habitually traveled through Trinity on a shortcut to the main shopping area, and a bread wagon made two stops at one end of the street. None of these drivers had been on the scene before the fire broke out. When they came the building was already ablaze.

The second-floor lights in number twenty-seven stayed on until past three one night and although we had spoken to every householder on the street we rechecked the woman who lived there on the basis that many people would lie rather than get involved in a police investigation. She said that she often watched late-late television but never had reason to look out of her windows while doing so. In any event the late movie on the Saturday had been a Jerry Lewis, and she hated Jerry Lewis. Only two men on the block worked a night shift that got them home after midnight, and neither worked Saturdays. The block seemed remarkably free of stop-out barflies and stop-up insomniacs, and our stake was running dry, when, at 2 A.M. on the third night, a cream Buick Skylark pulled up in front of number twenty-six and doused the lights.

We were parked to the south, also with lights off, and we waited ten minutes before approaching the car on foot and tapping on the driver's window. We could not see inside because the glass was fogged up. A young man with light-colored hair and wearing a turtleneck sweater rolled down the window after a few moments, and to his side we glimpsed a skirt coming down over naked thighs. A girl of about eighteen, very pretty, pushed herself into the far corner of the front seat.

"Excuse me," said Winters. "No need to fear. We're the fire marshals investigating last Saturday's fire at the apartment house. Since you're here so late tonight, we wondered if you happened to be in the area at the time of the fire."

"Yeah, we saw the fire," said the young man.

"It was terrible, those poor children being carried out." The girl was recovering her poise. She saw we were not interested in their lovemaking.

"You live here, miss?"

"One house back, twenty-four. We were saying goodnight. Didn't want to disturb my parents, you know." The girl was doing the talking. The man fumbled about with a cigarette. He was a yo-yo.

"What time were you here last Saturday?"

"About two, I think," said the girl.

"Did you see anyone in front of the apartment house then or later?"

"It was pretty dark," muttered the man. "We were only here a minute."

"There's a streetlight up ahead," I said. "Look, dammit, we believe that fire was set deliberately and we want to find the man who did it. Please, if you know anything, saw anything, Let's have it."

The girl moved to the center of the seat and put her face closer to the window. "I want you to catch that bastard too," she said, her voice rising. "We got here around two, maybe later, and started necking. There was a little van like electricians drive around in parked in front of the apartment that caught on fire. I remember it because there were no other cars between us and the van. A little while later we heard it start up and drive off. Then we saw this tremendous flash through the windows in the front doors and lights started popping on all over the place."

"What color was the van, miss?"

"Green," she said. "An icky green."

The girl later identified Frassa's van as the one she saw on the night of the fire, and this testimony, added to the other pieces, had him indicted for arson and homicide by a grand jury.

What happened to him after that I do not know; we were never called for a trial, and I assume he pleaded guilty to lesser charges, perhaps arson in the second degree and manslaughter, and drew ten to fifteen years.

I was no longer inclined to dwell on cases, and there was no time for it anyway. Fresh jobs were crowding in on us, two and three in one night. Winters was into an Indianapolis Speedway act across the highways of Brooklyn and Queens.

Bound for Far Rockaway one night, with me dozing in the passenger seat, he took the Cross Bay exit off the Shore Parkway at seventy-five miles an hour and went into a lurching skid. My eyes popped open to see a bordering cyclone fence rushing to meet us.

"Oh, no! Oh, no!" Winters' cry made my heart stop beating. He had never lost control of a situation in his entire life.

At the last instant before death, he hit the gas with one foot and the brake with the other. We skimmed off the fence and slewed back onto the paved ramp.

"It's okay," Winters said brightly. "Just a little action to keep you alert."

Chapter 9

Murder in
Rego Park

THAT night of the cyclone fence we were answering a call from the 47th Battalion, via Queens dispatch, for an investigation of five separate fires in a single apartment on Waterman Street, Far Rockaway. Firemen had tried to detain a drunken woman at the scene, but she had laid into them with a snow shovel and taken off.

"Say again," I had said when we received the alert somewhere in central Brooklyn. "Five fires? A snow shovel?"

"Confirm. Acknowledge response Car 50." Dispatchers were a laconic breed.

The apartment was actually the ground floor of an ancient, peeling two-family house, and waiting for us at the top of the front stairs was a distraught man with an emaciated body and a weather-worn face. His name, he said, trying to keep his voice steady, was Colin Corcoran. His wife, Daphne, had gone stark, raving mad, set fire to the house, and gone running home to her mother.

Corcoran led us inside the musty-smelling house and showed us the drenched, charred remnants of a fire at the foot of the mattress in the master bedroom, another in a mattress half pulled to the floor in a back bedroom, a third in a stuffed chair in the living room, a fourth against the curtains of the same living room, and a fifth on the ironing board cover in the kitchen.

Corcoran forced himself to hold back the sordid narrative of this terrible night until we got to the kitchen. Then he let loose.

"Goddam woman nearly burned me to death. Another minute and I'd be cremated. She went out of her head with the sauce, been drinking all night. A whole goddam bottle of vodka and into another. I went to bed to get away from her. I get to sleep and next thing I know my feet are boiling. I wake up and there's the goddam bed on fire, the bed I'm sleeping on. I run out into the living room and there's the drapes on fire. I run into the kitchen to get water and there's the goddam ironing board on fire. She comes in the other door then, screaming like a banshee, and starts picking up cups off the shelf and throwing them at me. She's loco, gone completely crazy. I run out the front door and go to the neighbor's and he calls the fire department. Jesus, those firemen are here in a minute. They're getting out the hoses and there's Daphne on the porch cursing at them like a sailor. They go past her with the hoses and when everything's out they try to put her in one of the cars. 'Easy, lady,' they're saying. 'Easy lady.' She gives them easy lady. She picks up a snow shovel from the side of the steps and whacks it across one guy's back. Then she runs off down the street, going lickety-split for her mother's house . . ."

Corcoran ran the fingers of both hands through his thin brown hair. There was spittle at the corner of his mouth.

"Whoa back there," said Winters. "We get the picture. We can see it. You must be advised, however, that you're the only one that can bring charges against your wife and you're the husband. You can't be forced to do this, and although you're understandably upset now you may feel differently in the morning."

Corcoran had lit a cigarette while Winters was talking

and he was calmer. "I'll feel worse in the morning. No, this can't go on. She's dangerous. I'll sign a statement or testify in court, whatever you need, and then maybe we can get her committed. She needs medical attention, psychiatric help."

The three of us got into the car and drove east for six blocks, at Corcoran's direction, to a big green-shingled house that had been converted into four apartments. Daphne Corcoran's mother, a frail, white-faced woman, opened the door to her second-floor apartment as soon as she heard us on the stairs.

As Corcoran stepped into the living room, Winters and me tagging behind, a small-framed woman in a stained blue housecoat and yellow hair flying over her face jumped up off the couch, yelling at the top of her voice. "You sonofabitch! I'll fix you!"

The woman did not give us a glance. She ran into the kitchen and emerged brandishing a long-handled fiber broom. Her husband backed through a door into the dining room and went to the far side of the table. "Daphne, please!"

With one mighty sweep of the broom the wife sent a heavy brown vase crashing off the table. Then she attacked the glass chandelier overhead, banging at it in a frenzy until the room was littered with broken glass. She threw the broom aside and bent to the floor. I watched, dumfounded, as she slashed at her left wrist with a sliver of glass, making the blood spurt, and jabbed then at her right wrist.

I bounded across the room and caught her at the back in a bear-hug. She threw off the grip, this one hundred pound woman, as if I were a child. Turning, she started to pummel me in the face with her bloody fists. Out of the corner of my eye I saw Corcoran escaping into the living room. I could not see Winters anywhere.

Roughly, I spun the mad woman around by the arms. She crouched over, lifting me off the ground, and charged around the room with me on her back. I was getting desperate. She was as strong as an ox. I saw Winters then, down on the floor, grappling her about the legs. She came down like a stunned animal. Winters pulled off his belt and strapped her legs together. I used my belt as a tourniquet on her left upper arm. Blood was pumping out of her gashed wrist.

"We better call an ambulance." Corcoran had appeared again. His wife had gone into a dead faint.

"No time," said Winters. "Into the car with her."

We carried her down the stairs and Winters propped her in the back seat while I dived for the wheel. Corcoran hopped in the front seat with me.

I gunned the car away from the curb, lights and sirens going, and grabbed for the radio.

"Car 50 to Queens dispatch. Notify Peninsula Hospital we're bringing in a woman with serious cuts on the wrists. ETA ten minutes."

"Kay, Car 50."

I made it to the emergency room in seven minutes. Corcoran was weeping during the ride and he kept squirming about to look at his wife. "To come to this after eighteen years," he moaned. "That's how long we've been married, marshal. Eighteen years."

His wife was coming out of her swoon when we arrived, but she was limp with exhaustion and sat quietly while two young interns cleaned her hands and wrists. She had not severed the arteries, and once the doctors had sutured the wounds we were able to take her to the 100 Precinct and book her. (She was made comfortable in a detention room and at the court hearing next morning was remanded for psychiatric evaluation. We heard no more of the Corcorans and assumed that Daphne had been admitted to a mental institution.)

Heading out of Rockaway in the predawn, we drove slowly over the Cross Bay Bridge and saw the warm glow of fishermen's lanterns along the walkway on both sides. Some of the men busied themselves cutting bait and repairing tackle, others were resting their elbows on the balustrades, gazing off into the inky waters of Beach Channel, their poles angled at their sides.

"Daphne Corcoran is trying to burn her husband to death in Rockaway and these people are out here fishing," I said to Winters. I was bone tired. My suit was saturated with blood, and I could feel blood matting in my hair.

"Yeah," said Winters. "Life goes on."

"It's good to be reminded. If you let yourself get too troubled by the ugly stuff we get involved in, you can't function."

A grisly picture came into my mind—the body of a Chinese girl, a prostitute, about twenty-nine years old, naked on her back in a cheap apartment on Manhattan's West 47th Street. She had been knocked unconscious and the apartment set afire by one of her customers. I remembered rolling the girl over. The skin had been burned off her back and buttocks, exposing bone. Later, at the City Morgue they had examined her on a steel table and then, when it was over, dumped her to the floor. Just dropped her, like an animal carcass. Her body hit the tiles with a dull thud, and her head flopped to one side. Then they opened up a plastic body bag and stuffed her in—human garbage for burial in a pauper's paddock.

"Jesus, I feel lousy," I said to Winters. "People can be so goddam inhuman."

"Forget the Corcoran woman, for crissakes. She's probably going through menopause and that's what set her off."

"She started me thinking about something else, a Chinese girl I saw roasted. Some people do such horrible things to each other you wonder whether they've reverted back to some inhuman stage. You know, not intentionally, but whether their minds have twisted up somehow and made them that way. Look at tonight. Here's a husband a woman has been married to for eighteen years and she sets his bed on fire while he's sleeping on it. It doesn't take much not to wake up, just a little carbon monoxide. Or if the flames had gotten to him he could have had ex-

cruciating pains from the burns. Could she have that
much anger for her husband? I say there's gotta be more
to it. The bed has to mean something. A person's bed is
security from birth. A juvenile sets fire to his bed when
he's punished by his parents and he's crying out for a
change in his life. Daphne Corcoran not only sets the bed
on fire, but her husband's sleeping in it. What torment
must have been going through her mind . . ."

"We'll never know." Winters was thoughtful now too as
he drove back around the horn. The harbor lights were
coming up on the left.

"That's the part that makes me so angry about tonight,"
I said. "The not being able to understand."

I rolled down the window and let the cold air blow on
my face.

Every good criminal investigator is a student of people.
He begins a case suspecting everybody, accepting nothing at face value, and progresses according to his capacity
for hard work and the strength and daring of his hunches.
Intuition, as Albert Einstein said, is as vital a force for discovery as imagination and expert toil.

My last case with Phil Winters became an exercise in
hunch play. It started around four o'clock on an April
morning when we were crossing from Brooklyn to
Queens, over the Kosciusko Bridge.

A second alarm sent us speeding off to a fatal fire in a
six-family house on Spencer Street, in Rego Park. We
were not much behind the fire apparatus, but ladder men
had already evacuated the building and had the fire
knocked down on all but the top, third floor. The place
was old, predominantly wood, and it had gone up like a
tinderbox. Three bodies had been removed from the third
floor and identified at the scene: Gloria Maxwell, a
woman of twenty-two, her infant son Randy, and a thirty-
eight-year-old spinster named Eugenie Clutterbuck. The
Maxwells had lived in the back left apartment, the Clutterbuck woman next door.

Firemen told us the shingled front of the building was a

sheet of flame when they arrived, apparently spreading from a sizable blaze in the first-floor hallway. We went into the devastated hall and confirmed that the blaze started over an area of several square feet at the center. Flames had raged up the stairwell as well as being sucked out the front door. Absence of electrical wiring in the area or anything else to trigger an accidental fire convinced us that it was arson. The shallow charring on the hall floor ruled out the use of flammable liquids and turned our thinking away from a premeditated revenge strike against someone in the building or a hired torch.

We spent a long time poking around in the ashes and debris for clues and came up with an assortment of blackened cans, bottles, fragments of materials, and several small circular springs of the type used in box mattresses and seat cushions for couches. Unless stacks of garbage were kept in the hall, I reasoned, somebody must have transported it there from an apartment or from outside. And very likely this trash was set alight.

I checked the battered metal garbage cans in a culvert by the front steps. Some were full, others empty, nothing deduced. A short distance down the street I saw a pile of old furniture left by the curb for city pick-up. The main piece was a motheaten coffee-colored couch. Two of the heavy seat pillows were missing. Several boxes of old drapes, blankets, cans, and bottles were stacked to one side of the couch. I tore open the remaining pillow, removed a spring, and took it back to the hallway for comparison. A perfect match.

"Maybe I'm leaping to conclusions," I told Winters, "but I think we have a dangerous psychopath around here, a person capable of making at least two trips to the sidewalk to get enough stuff to light a bonfire under a building of sleeping families."

Winters looked at me quizzically. "And that means, Sherlock, there must have been something or someone in this building to set him or her off. I can't see it as a random shot."

We circulated among the evacuated tenants and the bystanders outside, gathering names and scraps of basic information. The only person who had escaped the third floor was Gus Maxwell, whose wife and baby had died, and I asked around for him.

A hand clutched at my sleeve. "Officer, I have Gus in my car." The voice came from a rugged, black-bearded man of about forty. He had deep wrinkles under his eyes and an ugly strawberry mark on his right cheek. "I'm Mervyn Schlatter, a friend of Gus's."

Schlatter led me to an old black Chevy. A man sat in the front seat crying.

"I'm sorry, Mr. Maxwell," I said, bending toward the open window. "May I ask you a few questions? I'm from the fire marshal's office and I'm afraid I must talk to you now."

Maxwell raised a sooty, tear-streaked face. "My wife and baby were in there. They're dead."

I crossed to the driver's side and got inside beside Maxwell. I was irritated to see Schlatter climbing into the back seat.

"Sorry, Mr. Schlatter, the questioning has to be private."

"It's my car, goddam it. I want to stay."

"I don't give a shit, Mr. Schlatter, about whether it's your car. Give us a break. Get out."

He backed off about twenty feet and stopped there staring at us.

I ignored him and began talking to Maxwell. He wore only a pair of pants. His torso was thick-muscled and smudged with soot. Tattoos were inked on his biceps. They appeared to be sea serpents, though I could not be sure in that light. He had tousled dark hair, full, over the ears, and his face was strong and handsome. At most he was twenty-eight.

"Tell me what you can, Gus," I said quietly.

He sobbed as he began speaking, but his voice became stronger as he went along. Every detail came out without any prompting from me. The heat of the fire awakened him, he said. He heard the crackling of flames, smelled smoke, and dashed from the bedroom through the apartment toward the hall door. Flames were shooting under the door and curling up on the inside. The heat blast was almost overpowering. He ran back to the bedroom, shouting at his wife Gloria to grab the baby and come to the fire escape, which ran down the building on the living room side. There were two windows in the living room and in his panic he threw open the wrong one. The fire escape platform was some feet from this window. He climbed on the sill, leaped for the railing, missed, clutched the angle iron holding the steel ladder to the side of the building, and managed to haul himself up to the platform. There were potted plants on the wide sill and he used one to smash out the glass in the window. Smoke and flames filled the apartment. He heard Gloria screaming and he lay on the platform, thrusting his arm into the room and calling for her. The screams suddenly stopped. Maxwell realized his wife was unconscious and probably dead. Down the fire escape he went like a madman, intending to run to the front of the apartment house and try a rescue from the main stairs. That was when he ran into Mervyn Schlatter, a close, long-time friend. They approached the entrance together but were met by a wall of fire. Schlatter comforted his friend and took him to his car.

At the finish of the story I cleared my throat and apologetically asked if he had any enemies who might want to kill him. In my opinion, I told him, the fire was set deliberately.

"This is where I was raised," he said. "In this district. I have nothing but friends here."

"Well, is there anything more I should know about you that might help the inquiry?"

"You'll find out," he said slowly, "that I have a record. I did eight months in Queens County for fencing stolen watches. I been out for two weeks."

I left him with his head dropped forward on his chest

and his hands clasped together so tightly that the finger-nails showed bloodless white.

Winters had nothing remarkable to report from his interviews. I filled him in on the Maxwell story as we picked our way up what was left of the stairs to the third floor. Maxwell's wife and child had been found dead on the kitchen floor, which meant she had gone past the living room. This was not too significant. She would have been disoriented by the smoke and by fear. We poked our heads out of the open window through which Maxwell had made his escape. The fire escape railing was a good six feet away.

"I couldn't make that leap in a fit," said Winters.

"Nor me. You think the guy was bullshitting?"

"He does have a record."

"Yeah." I was thinking of Maxwell's honest grief, and also of his muscles. "Maybe it could be done. You gotta remember the guy had a fire licking at his backside."

Schlatter was sitting in the Chevy with Maxwell when we came down. Winters escorted Maxwell to our car and I paused to make peace with the bearded one.

"Rules are rules, Merv. My talk with your pal had to be confidential."

"I thought I could help." Schlatter was relaxed and friendly now.

"You can. You were one of the first on the scene."

"It was pure chance," he said. "I was drinking late at a bar on Montgomery and was driving past here on my way home when I saw the fire. Jesus, what an inferno. Soon as I got out of the car I saw Gus running around the side of the building looking like he'd been stoking coal for a week. He wanted to go back into the house. I stopped him; it would have been suicide. At that point the fire engines started arriving."

"Where do you live, Merv?"

"Over on Seventeenth, number eighty-nine."

"Isn't that the other way? I mean, we're east of Montgomery. Seventeenth is west."

"I wasn't sleepy. I wanted to drive around a while."

Schlatter accompanied us to the precinct, where both he and Maxwell made formal statements to the homicide detectives. We had ruled the fire a definite arson and were sharing the investigation with the local talent. Maxwell was known to the police—a tough neighborhood kid who grew up to run with a shady crew. They put him through three hours of tough questioning without breaking the story he told me in the car.

Late in the afternoon we released both men. Winters and I split for home, arranging to meet at the precinct at nine the next morning. I thought a lot about Schlatter that night. He had said at the precinct that he was an electrician by trade, had a wife, three children. Why would he be out drinking by himself until four in the morning on a Wednesday? And if he had worked all the previous day, as he said he had, it was odd that he should be that wide awake that he would want to drive around in his car, a drive that coincidentally took him past his close buddy's house. If he was disturbed about something he did not show it at the fire, apart from that one incident when he was reluctant to let me question Maxwell alone. He was miffed, however, not upset. He should have shown more distress. These people were his friends. In fact, I heard from the police that it was Schlatter who made the first identifications of the three fire victims. For any normal person that would be a heart-rending experience, yet Schlatter did not appear saddened.

On the way to the precinct house in the morning I checked on Flynn's Bar, where Schlatter said he had been imbibing. Flynn, the man himself, confirmed Schlatter's presence on the Tuesday night. He was the last to leave when the bar closed at one o'clock.

"One o'clock?" I stopped him right there.

"Yeah, one. I was off color and tending bar by myself. There were only a handful of men here. They didn't mind me closing early."

I told homicide about this flaw in Schlatter's story and they agreed that Winters and I should pick him up. His wife answered the door to the duplex on Seventeenth. She was a tall woman with white-gold hair and she kept her house in mint condition. We waited in the parlor while she called Schlatter, who was asleep upstairs.

"Merv told me he had answered all your questions yesterday," she said. Her eyes were bloodshot and her cheeks drawn as if she was under mental or physical strain.

"It's strictly routine, Mrs. Schlatter."

"That's what they're always saying on television," she said in a faraway voice. We did not push the conversation.

Schlatter came down the stairs briskly. He was dressed neatly in a leather jacket and brown checked trousers, and he had taken time to comb his hair and smooth his beard. His strawberry mark looked lumpy and beetroot red in the morning light. He kissed his wife goodbye brightly and waved us through the front door with a courteous "after you, gentlemen."

A group of homicide detectives took turns with him for most of the morning in the squad room annex. He brushed away the discrepancy in his story about Flynn's Bar, saying that he had actually been to several bars that night and would have said so but did not want to give the impression that he had been on a binge.

I canvassed Spencer Street with Winters and we scored a female witness, a stout, middle-aged woman, who said she had been walking her dog at the end of the block on the morning of the fire and thought she had seen a man throw a burning stick into the Maxwell apartment house, then get in a car and drive away. She was extremely vague. A defense attorney could cut her to ribbons. Of only one thing was she certain. The man she had seen was heavily bearded.

Chief Canty was with the precinct commander when we returned about noon. "We've come up empty," Canty told us. "Schlatter won't budge and we have no real evidence against him."

"May I talk to him by myself for a minute?" I asked.

"That is, if the officers don't mind and it's okay with Phil."

"You're running the case, John," said Winters.

"Sure thing, marshal," chimed in the commander, though he looked daggers at Winters for his comment.

With the homicide men cleared out of the small annex room, I sat and talked and smoked cigarettes with Schlatter for half an hour without attempting interrogation. I told him about my job, the long duties we pulled, and he said he was no stranger to weird hours himself. He was a freelance electrician, available night and day. I kept leading him. Sometimes, he said, he wondered if suburban women called tradesmen like him for repairs or romance. Midmorning was the key time for them. They'd open the door in a transparent nightgown and hang around while he fixed an ancient toaster or replaced tubes in a television set.

"I could get laid every day if I wanted it," he boasted. "Funny thing is that I turn down all this terrific pussy and I go home and get the ice treatment from my own wife. That's a laugh, isn't it, a real laugh."

He stopped abruptly. His voice had risen sharply in bitterness and his eyes were glistening. The strawberry mark, the size of a half-dollar, glowed fiercely.

"Merv," I said gently, leaning forward. "I know you set the fire on Spencer Street. Maybe you didn't mean to do it, but you went ahead and put that rubbish in the hall and you made a big fire. It was a mistake, Merv. We all make mistakes and I've made plenty myself, but this was a mistake that shows you need somebody to help you. Now is the time to get that help. This will be the last chance, Merv. I want you to tell me the whole story from the beginning to end."

Schlatter heard me out, his lips trembling. Then he put his head on the table and cried, great wracking sobs shaking his whole body. The door behind me opened and I spun around to see Canty's red face glaring in accusation, and craning behind him the heads of two homicide men. I waved them away with my hand. The door closed.

"Let's stop playing games with each other, Merv," I persisted. "Tell me the story. It's the only way I can help you."

Schlatter lifted his head off the table a few inches and I saw wet marks on the bare wood. He stayed with his head down, talking in a hoarse, continuous voice.

"I got up at midnight on Tuesday and got dressed and went to Flynn's. There was no action there. Flynn was sick. I drove up to a bar and grill on Fourth and had a hamburger and bottle of beer. The waitress is a friend of mine, a sort of a friend. Her name is Terry. I had brought two wigs from home, a blonde one with curls and a beautiful chestnut fall that went to the shoulders. I make wigs, you see. Real human hair. It's my hobby. I told Terry she could have the chestnut wig if she let me drive her home. She told me to get lost. She called me a creep. I tried another place then, the Sand Bar, over by the bridge. This has a piano, everything. I took a table and ordered a beer. The cocktail waitresses there wear short skirts and net stockings. Real cute. It turned me on, I'll admit that. My waitress had her boobs pushing out of the top of her dress. 'Steady, Merv,' I told myself. 'Steady.' I tipped the girl five dollars on the first drink. When she came back the second time I showed her the wigs and said she could have her pick if she met me after closing. She giggled, you know, and said she would think about it. Her tits were rising like big ice creams. They were driving me crazy. I had three drinks and each time I tipped her five dollars. Just before closing the bartender came and told me to leave 'cause I was making a nuisance. Bastard. I was going to smash him in the face, but he got out the brass knuckles, you know, and slipped them over his hand. I drove around for a little bit after that and I got to thinking about Genie. Genie Clutterbuck, the woman who lives, who lived, next door to Gus and Gloria. I'd met her when I was over at Gus's. I went over to her place. It was all dark, and I knocked on her door for a long time. She never answered me. I went downstairs again to the car. When I opened the door I saw this red emergency flare on the floor. It must have rolled out from where I kept it under the seat. I guess I was feeling pretty mad, or frustrated, or something, because I lit it and threw it through the door of Genie's apartment house. The door was open. I had left it open. I drove away and was going to go home but something made me go around the block and come back. I saw a faint red glow on the floor in the hall. Jesus, it was nothing. I got out of the car again and noticed the pile of old furniture on the sidewalk. There was a couch there and I grabbed two cushions and took them in and put them on the flare. Then I got a bag and a box of garbage from the same pile and dumped these onto the cushions. Goddam it, I wanted a fire that would wake up the whole apartment house. I drove around the block again and when I got back this time I couldn't believe it. The flames were shooting out of the second floor windows. See, it was my fault, and yet it wasn't my fault. And it turned out good in one way 'cause I found Gus, and Gus needed me."

Schlatter raised his head. "That's the God's honest truth."

"It had to come out, Merv," I said. "Now I want you to go a step further. I'm going to call in those detectives and I want you to repeat the story to them. The details will be taken down in writing and you'll be required to sign them. Okay?"

"Sure thing, marshal." Schlatter was composed again, even pleased, I thought, at the prospect of staying at the center of attention.

I felt a nauseous excitement when I opened the door to the squad room and announced to Canty, the precinct commander, and three homicide men that Schlatter had made a full admission and would repeat it for them and sign a confession.

I glanced at the surprised faces and fixed on a beefy detective nearest the hall door. "Say, how about getting him a cup of coffee and a fresh pack of cigarettes," I said. "He smokes Viceroy." It was a petty gesture of triumph, straight from homicide's cock-of-the-walk rulebook. The cop smiled and hopped to it.

Detectives made the arrest and handled the arraignment. The case went quickly through the courts and Schlatter was sentenced to twenty-five years to life.

Credit for the arrest was marked to homicide, with scant mention in the police records or the newspapers of the role of the fire investigation division. I had expected this to happen because of the absurd rivalry between the police and fire departments in New York, but it was impossible not to feel bitter, and difficult afterwards to work a case with police without the suspicion that they would step in at the eleventh hour and steal all the thunder. The rivalry persists to this day. It sabotages the spirit of teamwork in arson investigation and doubtless is a source of fine amusement to the arsonists.

Chapter 10

On My Own

COLONY Hill is a pleasant old neighborhood in the east Bronx—clean streets of one-family homes, sections of masonry row housing, a scattering of splendid brickfronts, grand churches and delightful parkland. The 'sixties had seen an extensive movement of black families into the area and several black associations had been organized to improve the city services, notably the schooling, expand the sports facilities, and erase the last blocks of tumbledown wooden tenements for new housing. In my experience, totally distorted by the Black Panthers, black groups in New York were querulous and belligerent, and I was stupidly slow to respond to complaints from the Colony Hill Citizen's Action League that vandals were overturning tombstones and breaking windows and throwing burning newspaper darts into the administration building at the local cemetery. The cemetery was on a five-acre plot at the edge of the city, screened by trees, and poorly lighted. The administration building was right at its center—a small, square, two-story brick construction that was not staffed at night or weekends.

"The only way we could catch the offenders would be to stake out the cemetery, possibly for weeks, and we don't have the manpower," I told Cyrus P. Strong, the league president, when I caught his telephone call to the division office.

"I thought fire marshals were responsible for investigating cases of arson," said Strong.

"Absolutely," I replied, "but all you have is a few charred pieces of newspaper against a whole pattern of vandalism, and vandalism is rampant in every single neighborhood in the city."

"Thanks for nothing," said Strong, and hung up the phone.

Three nights later a Molotov cocktail was lobbed through the second-floor window of the building, demolishing the upper section and roof.

I went straight from home next morning to examine the fire, then visited Strong, an oak of a man with grizzled gray hair and fierce reproach in his eyes. Strong's group had taken over the administration of the cemetery and ran it as a separate nonprofit community association. I apologized for not making at least some inquiries about the previous minor incidents and promised to find the arsonists.

"If there's the slightest chance, I'll make sure you get full restitution for the damages," I added.

"Just find the punks who threw the Molotov," said Strong. "We want them brought to account. And if you want a place to start looking try Bush Haven, the place to the west. A caretaker saw white kids from there splashing paint on tombstones. Know what they wrote on one of them? 'Here lies a good nigger.' "

Racism is not one of my faults. I could never understand the basis for it. But I could recognize a potentially ugly racist situation here: predominantly white arson investigation division ignores appeals from black community, wanton acts of arson are committed against community, division "fails" to find perpetrators. Chief Canty saw it too, when I reported in, and ordered me to stay with the case by myself until I had it solved.

I took Strong's advice and began at the police station in Bush Haven, a white, working class district. The officers had no definite leads on the vandalism in the cemetery, but they had the names and addresses of members of the local street gang and I went through these one by one until I found a teenager who had heard that the fire was the work of Paul DeGaris and Ken Duncan.

DeGaris, aged sixteen, readily confessed to the arson in the presence of his father and stepmother. The father was a chunky construction worker with an open face who kept clenching and unclenching his fists while I described the destruction caused by the fire.

"Oh, God, no!" he groaned when I had finished. "You're gonna be locked up for this, Paul. Why did you do it? Sneaking around in the dark, throwing fire-bombs. It's cowards' work."

"I dunno," said the boy dully. "The boys are always going over to the nigger cemetery for some fun. Last night Ken and me took a Molotov and just decided to throw it into the building on the spur of the moment."

I told the father that his son would be arrested when I came back with Duncan.

The Duncan house was an old wooden place in a street off the Bush Haven shopping plaza. A woman with dry blonde hair and fatigue around her eyes opened the door. She said she was Ken's mother and I said I was a fire marshal anxious to talk to him about a fire.

She closed the door quietly behind her and led me by the elbow to the lawn out front. Her hand was shaking. "What's the trouble?"

"I have information that your son and Paul DeGaris set a fire in a cemetery building at Colony Hill. There's no doubt. Paul has admitted it. The black community is up in arms about it and there is no way I am going to take this thing lightly."

Tears shimmered at the corners of the woman's eyes. The purple discoloration underneath made her appear quite ill. "Kenny's been going wild lately, I won't deny it. But never a policeman at the door." Her voice was breaking. "We are going through terrible times in our family. My husband, you see, is dying. He's . . . the doctor's given him three months. Gordon has lung cancer, inoperable. He's getting weaker and weaker. His body is getting weaker. His mind is good yet. Kenny's his whole life. He adores the boy. If you take him away you will kill my husband."

"I'm sorry, Mrs. Duncan. I don't know what I can do. Please let me talk to Ken."

She went inside and sent the boy out. He was in jeans and a green football jacket with 67 on the arm in yellow numbers; a halfback, I guessed.

"My mother told me why you're here, sir," he said. "I threw the Molotov, all right, but I didn't mean to burn the building down. It's too late to be sorry, I suppose."

"Damn right it is. Why would you do a thing like that?"

"No reason," he said, "except that we wanted to throw a Molotov and see what would happen."

The mother joined us then. "I have a suggestion, marshal. Kenny is about to turn eighteen and he's got his enlistment papers into the Army. He'll get straightened out there. That's what his father says: 'Every boy should have a hitch in the Army; makes a man of them.' Please, marshal. I'll take the responsibility and I'll repay every penny of the damage. Overlook this one mistake. I'm begging you for my husband's sake."

"Ma'am, your son's committed a crime, arson in the fourth degree, at least. He could be the Mayor of New York and still have to pay the penalty."

The woman wept then, soundlessly, tears pouring down her cheeks. I got the name of the husband's doctor and said I would return the next day.

From home, I telephoned the DeGaris boy's father and told him to keep his son close by for a day or two while I made more inquiries. I reached the Duncan family physician in his offices and he confirmed that Gordon Duncan was dying inch by inch of cancer. He gave him two months at most.

The boy could be arraigned and released on bail, I thought, and the case delayed until after his father's death. Even in the remote chance that this could be kept from the father, however, the mother was at the end of her tether and would surely go into serious nervous breakdown. Furthermore, the son would be scrubbed by the Army, sent to jail, and the mother might never recover from this. But the boy had done a dreadful thing and I could not be sure he would not do it again. Had the tragedy with his father made him strike out against the world, or was he a young savage in need of restraint? I did not know the answers. As for DeGaris, whose fate rested with that of Duncan's, he was a follower, not a leader, and he had a healthy father who seemed to have the right set of values. Some kind of probation might serve here. Revolving around and around in my mind with these considerations was the sense of straight duty I had as a fire marshal to arrest arsonists, whatever the extenuating circumstances, and also the extraordinary pressures to make those arrests. The folks in Colony Hill had rightfully demanded that the guilty be punished. The fact that the guilty were colored white made this the more imperative, and the more certain that there could be serious reprisals if I did not go by the book.

I checked with Bush Haven police on Ken Duncan's prior record. Nothing. Why did he do it? I remembered an incident from my own teenage. I was about fourteen and I was crossing the old South Beach railroad tracks on the trestle bridge—walking outside the guard rails to show off. A long freight train rumbled underneath, shaking the bridge, and still I kept going. Halfway across a tabby cat perched on the timbers in front of me, licking its fur. I swooped down, catching the cat under the belly, and dropped him onto the brown, rusted roof of a car passing below. He fell crouched on his paws, and rode off into the distance. Why did I do something as thoughtlessly dangerous and cruel as that? And was it anywhere near the category of Ken Duncan's fire-bomb anyway?

In the morning I called Chief Canty and related the details. The only possible way out, he said, was to get Cyrus Strong and his association to withdraw charges, forgive the arson, though he doubted that human charity could be that generous and whether I could make a stout enough case to ask it.

"You must be made aware also, John, that this is a hot potato and every action you take from here on in this case is entirely on your head," said Canty. "You are departing from standard procedure and if you foul up you're going to have to take the consequences.."

I made the long drive to Cyrus Strong's stationery store, struggling to keep my nerve up. For the second time that day, in Strong's back office, I went over the story from beginning to end. He did not offer any comment at the finish, merely asked me to browse through the store while he made some telephone calls. It would have been within his right, I realized, to make an official complaint that could cost me my badge. In fact, I feared it. Two white boys torch a black building and a white cop comes by and attempts to get them off the hook. Strong must have thought: what if the situation were reversed and black kids had fire-bombed a white building? Would they have anyone appealing for them to go free? I preferred to think that the answer would be yes—if the cop was me.

Within twenty minutes Strong hailed me back into his office. "You are a compassionate man, marshal," he said. He had lit a huge cigar and the room smelled marvelous. "We have decided to let you handle this matter according to your own conscience. We make the conditions that the desecration of our cemetery should cease as of this minute, that you ensure the Duncan boy goes into the Army, that you devise some suitable lesson for the DeGaris boy so that he never again is tempted to damage other people's property. We have talked over the matter of restitution and want none. The building was insured and we have the funds to replace it with something more sacred for the purpose of laying brothers and sisters to final rest."

I sat there for some moments looking at Cyrus P. Strong, trying to find the right words. They did not come.

"I don't know whether I should say this or not, Mr. Strong, but I will anyway. My impression had been that all black community leaders were militant and vindictive. I was totally wrong. That is the lesson I have learned and I am grateful to have learned it. You, sir, are a wonderful person."

Strong put the cigar in his mouth and grinned around it. "Get outa here, marshal. You got a busy day."

Mrs. Duncan wept again when I told her, outside on the lawn, that her son would not be arrested if he could prove that he was entering the Army. She ran inside and came out with Ken in tow and copies of the enlistment forms. I ordered the boy to take off, there and then, to spread the word among his friends that the Colony Hill Cemetery would be staked out and if anybody so much as touched the fence I would personally beat them to a pulp. Furthermore, on threat of changing my mind about the arrest, I warned him that if Paul DeGaris ever asked if he was paying back a weekly amount to compensate for the destruction they caused, he was to give it the affirmative.

At the DeGaris house I found the father home, as I had expected. He was upset, torn between paternal loyalty and a conviction, clear at our first meeting, that such willful wrongdoing deserved strict punishment. I gave him the facts and said that Paul, though in high school, must find a part-time job and pay twenty dollars a week to the father for a period of two years on the understanding that this was restitution.

Every night for two weeks after that, from eight until midnight, I parked alone on the Bush Haven border of the cemetery. I saw nothing, and when I called Strong he confirmed that the acts of vandalism had stopped. Gordon Duncan died November 4, 1970, nine weeks after the cemetery fire, and Ken Duncan went to boot camp on November 23 . . .

Three years later, out of the blue, I had a postcard from young Duncan. He was serving in West Germany and intended to sign on for another hitch. I called DeGaris that night. Paul, he said, had paid him two thousand dollars. The father had returned the money to his son one year ago and explained the very special hand of justice that had been extended. The boy had applied the money to his fees for his second year at a New England college.

I did not have to check the cemetery. Our regular Bronx teams said they could see the new administration building every time they drove down Lafayette Avenue. It was a futuristic creation in redwood and glass, a fine chapel in the front of the structure, the offices tucked away in the back.

The Duncan case was one of many that kept the calluses from forming around my heart. Qualities like kindness and contrition, I discovered, were very much alive and well in the city.

In February of 1971, when I was working Manhattan with Sam Westerling, we answered a 2:45 A.M. call to Sheridan College, off Bleecker Street, in Greenwich Village. Westerling knew it as a small, expensive girl's college of strict academic standards. The curriculum was heavy on the sciences. Sheridan graduated a considerable number of premedical students.

The college was small all right—two double-front, six-story white stucco buildings side by side, one housing the school itself and the other the dormitories. The matron for the living quarters, Miss Ruth Jacoby, a pretty, pony-tailed woman in her early thirties, was waiting for us at the main door and took us along a passageway to an elevator. At the fourth floor we stepped out into a square carpeted area that had the stairs to the left and wide, glass-windowed swing doors to the right. The doors led to a corridor with maybe nine dormitory rooms on either side, much like a hotel arrangement. The matron said the four other residential floors were identical. There were 173 dormitory students and some 160-odd "day girls."

A few paces inside the swing doors we saw the reason for our summons. The first of two wood-paneled telephone booths was gutted on the inside. The floor to it was a mess of wet, charred debris, and the most cursory examination indicated that torn telephone books had been piled on the floor and set alight. The fire caught the timber on both sides and burned on through the roof to scorch the plaster wall and ceiling above before it was extinguished. The adjoining booth was also severely damaged. Another five minutes and the girls would have been jumping from windows.

"It's what you people call an inside job," said Miss Jacoby grimly. "The fire would have started about two-fifteen and I can assure you there were no strangers in the building at that hour, and no cleaning staff. Regrettably, it must have been a student. I was asleep on the third floor when I heard the girls screaming. I rushed to the stairs and saw smoke everywhere and immediately called the Fire Department."

"Has anything like this ever happened before?" It was Westerling's question.

"We've had four other fires, very minor," said the matron. "We were foolish enough not to inform the Fire Department about them."

I thought of the 173 girls in the building, all suspects and doubtless all with the same alibi; they were in dreamland.

"Let's try and narrow this down if we can," I said, thinking aloud. "When did the fires start, matron?"

"We had the first shortly after the start of the fall semester and they've been occurring at the rate of about one a month since."

"That would more or less rule out the older students, if there were no fires in previous years. How many resident freshmen, or freshwomen?"

"We did not have fires before and there are forty-three freshmen." Ruth Jacoby was a no-nonsense woman. She would run a tight ship, I imagined, and she would have an acute antenna for potential troublemakers. I asked her if she had had difficulties with any of the freshmen girls, whether any were particularly mischievous, and she said no.

The matron understood that we should question the girls immediately and gave us the use of a small office room opposite the telephone booths. I asked her to send in the freshmen one by one, starting with the fourth floor.

A few minutes later the first girl appeared at the door and crossed to the interview chair in front of the desk. I used the desk chair. Westerling was still examining the telephone booths.

"Could you tell me how you first became aware of the fire?" I asked the question automatically. My mind was not on it, nor on the answer the girl started to give. She was a pert blonde of about nineteen and she sat very straight with the thin silk of her nightgown falling against her young breasts. I tried to rivet my eyes on my notebook but they would not stay there.

"You'll have to forgive me," I interrupted. "I've never conducted an inquiry in a girl's dormitory before and I'll need time to get used to things." I sucked in my breath. She was a gorgeous little child-woman. "Would you please go and put on a bathrobe and tell the other girls to do likewise."

With Westerling's help I got through a dozen or so girls during the next hour. None of them advanced us one inch toward solving the arson and we had to pack it in for a while because our witnesses, including Miss Jacoby, had fallen asleep in their rooms.

We talked over the case during a breakfast of waffles and syrup on Bleecker and decided we had to pay more attention to the timing of the fires. Either they were set by an incipient pyromaniac or there was a reason behind them. Pyromania was unlikely because they were different types of fires—two with books in telephone booths on different floors, one in the laundry room on the second floor, using sheets as fuel, and two on staircase landings using what had been described to us as wrapping paper and shopping bags.

On our return to the Sheridan dormitory house at nine we reestablished ourselves in the office on the fourth floor and instructed Miss Jacoby to get the exact dates and times of the other fires. We struck oil. They had occurred very late, after midnight, on the third Wednesday night of each month since October, up to and including the previous night, also a third Wednesday.

I toyed with that coincidence for a long time. One of the girls was so distressed on one particular night every month that she crept out of bed and made a fire. Maybe it had some weird connection to her menstrual cycle? No, it could not be *that* regular. She was obviously not trying to burn the building down out of hate for her classmates; the fires were too small. They were like diversionary fires. I asked Miss Jacoby if any of the girls had reported theft during the commotions. The answer, again, was no. Lodged in the back of my mind was a notion that whoever set the fires was subconsciously wanting to be found out. I brought that thought to the surface. Assuming the girl wanted to be discovered, she also wanted to be expelled, and the desire was at its peak on the third Wednesday night of each month. The obvious hit me then: the fires must relate to something that took place on third Thursdays, something that the girl detested. They were a pathetic attempt to stop Thursdays from happening at all.

"Miss Jacoby," I said, "this is purely a hunch but I believe our fire-setter is a very unhappy young lady who may feel out of place in this college. I think she may be a poor student and poorest of all at a subject that is taught on Thursdays."

Miss Jacoby shrugged prettily. "You're a long time out of college, marshal. There are ten different subjects taught on Thursdays."

"Don't give up on me. I took education courses at Fort Campbell, but nobody believes it. Are there tests given in any freshman subjects on Thursdays, specifically the third Thursday of every month?"

"It's the regular scheduling for the chemistry test." Now she was interested.

"Okay, let's have a list of the girls who are borderline or failing chemistry."

The matron disappeared down the elevator for half an hour. I told Westerling how I had come to my brilliant conclusion. He thought it was thin, and I had to agree.

Miss Jacoby returned with the names of four girls and said she would send them into us separately.

The first girl was my beauty with the silken breasts from the night before. She was in a brown corduroy skirt, well above the knee, and a Kelly green sweater that showed her every curve. Her boyfriend, whoever he was, must climb walls between dates.

I was stern. "We've finished with the games, miss. Tell us what really happened last night and no cute stuff."

She glanced around at the door, presumably to make sure we were alone, and leaned across the desk, talking in a murmur.

"Miss Jacoby would kill me if she found out. I was out with my boyfriend until two-thirty. Some of the girls have a key to the freight door in back and I had borrowed it to sneak in that way. When I arrived the place was in a terrible tizzy. Firemen were wandering around with axes, for crissakes. I went to my room and laid low until they told me you fire detectives were here. I know I lied to you and said I was sleeping when the girls began with the hysterics, but, honest Injun, I haven't the foggiest idea how the fire started."

I took the name of her boyfriend, although my intuition told me she was not the one. She had the world by the tail, this girl.

The second student was a chubby, effervescent, talkative type that seemed to eliminate her from the outset. I had not seen her the night before and she was desperate to get in on the excitement.

"I was in my room up on the fifth floor when I heard the girls screaming, fire! fire!" Her voice went up on the exclamations. I looked at Westerling and he bobbed his eyebrows up and down. "I opened my door and saw the girls were coming out of their rooms at the end of the hallway and going through the doors and down the stairs. Naturally, I followed them. The ruckus was coming from the fourth floor, here, and we saw smoke billowing out right into the lobby. Gee, I thought, that's funny. That's what Janet must have smelled."

"Janet?"

"Janet Sakowitz. See, you didn't ask me if I was asleep when I heard the girls screaming. I wasn't, and the reason was that Janet Sakowitz, who is not at all a friend of mine, knocked on my door five minutes before any of this happened. She lives in the room opposite mine. Anyway, she woke me up and said, 'Hey, Barbara, do you smell something funny?' I said, 'What funny?' and she said, 'Like smoke.' We went out into the hall and down to the elevator, sniffing the air, you know, and I couldn't smell anything. 'There's nothing here, Janet,' I said. And she says, 'I still think I smell smoke, but maybe I'm mistaken.' 'Jeez, Janet, it's after two o'clock in the morning,' I go, 'let's get back to bed.' And so that's what we did."

"Go on, Barbara." The narrative was taking forever, but it was intriguing me.

"Okay, so when there was really a big fire, and only five minutes later, I knew that Janet must have smelled smoke, and I thought, what a strange thing. See, Janet's asthmatic and she's not supposed to be able to smell anything. But she could smell smoke and I couldn't."

"Does Janet have a roommate?"

"No. That's why she came to my door. Most of the girls have roomies, and I'd sure as hell like one. Janet and me have a room by ourselves, and we'd double up except that she makes wheezing noises in her sleep, she says, and I don't much like her anyway."

Janet Sakowitz was waiting in the hall and we called her in next. She was a timid, wan girl with pimples on her chin and close-set eyes. She perched on the edge of my interview chair like a frightened bird, while, as with the others, I advised her of her right to remain silent and the rest of the required Miranda preliminaries.

"There's no need to be afraid, Janet," I said, composing an enormous lie. "The reason we're talking to you is that we're told you were using the telephone down here on the fourth floor just prior to the fire."

I was prepared for her to deny it. How many people make calls at two o'clock in the morning? "Yes, as a matter of fact I did," she said. "I called my mother."

"What time was that?"

"A little after midnight. I had to talk to her about something important."

"Why didn't you use one of the phones on the fifth floor?"

She stroked a wisp of light brown hair falling down the side of her cheek, probably a nervous habit from childhood. "They were both occupied."

"Janet, we have information you were down here much later than midnight. We also know that you were seen near the other fires in this building. Come on now, tell the truth."

The girl began weeping. Her breathing came in shallow, asthmatic gasps.

"Please, Janet, there's no need to cry. Nobody was hurt in the fires. The damage was small. Let's just talk about the first fire, the one in the laundry room. Gosh, that was so tiny it burned itself out."

"That was me," she sobbed. "I set the sheets alight in the laundry cart, but I didn't do the one last night."

"Never mind last night. Why did you set the laundry fire?"

"I was doing badly in chemistry and there was a test the next day and I knew I wouldn't pass it. That was going to make my father very upset. He's mad keen for me to get through Sheridan with straight A's and go on to medical school. I'll never make it; I don't even want to. But father's a physician, you see. A cardiologist. I made a fire thinking they'd cancel the test and give me more time to prepare. They didn't though. They thought the fire was an accident and didn't ask anybody any questions."

I led her through the next three fires, which she also admitted. She kept to her stubborn denial of burning the telephone booth, even though a previous fire had been similar.

"You're off the hook this time, Janet," I said, studying my watch. "It's past ten-thirty. They've started the chemistry test without you. I'm concerned about these fires because if one got out of control a lot of girls would be hurt, and you don't want that. You say you can't keep up here at Sheridan, and don't really want to, so wouldn't the intelligent thing be to make a clean breast of everything and tell your parents exactly how you feel."

"It would kill my father."

"No, it wouldn't kill your father. You can't lead a life he plots for you. He'd be the first to see that if he had all the facts."

"All right," she sighed. "I set fire to the telephone booth. I called my mother, as I said, and she bawled me out for waking her up about the same old thing—my chemistry. She told me to knuckle down for dad's sake. That was that. End of conversation. I stayed in there a long time, having a weep. Then I pulled the telephone books off the shelf, ripped them up as best I could, and set fire to them."

I felt dread instead of elation at the confession; I had no option but to book and fingerprint her, and take her to the Tombs and then criminal court. There was no provision in the law for more sympathetic treatment for disturbed, pimply-faced kids like Janet Sakowitz. Her acts were criminal, yet she was not. Expert family counseling would solve her problems. Incarceration might mean the ruin of her life. I could not extend the same charity as I had in the Duncan case because Janet had set her fires in an occupied building, a major felony. I pushed a pad and pencil toward her and asked if she would write a brief admission and sign it. Westerling, who had stood silently through the interview, volunteered to fetch coffee. I went to find Miss Jacoby in her ground-floor office.

"I'll contact the parents and make arrangements for her dismissal," she said after I gave her the story. "They live out on the Island."

"Good. Tell them to come to Manhattan and meet me in Part One of the criminal court building on Centre Street. If they ask around somebody will show them the arraignment room. Warn them that the case may go over into night court."

On the way to the precinct, a few blocks from the college, Janet commenced to cry again and her skinny body trembled all over.

"Please, John" she sobbed. "I won't set any more fires. Please don't send me to jail."

"It'll be okay, Janet. I'll tell the judge you've been under terrific strain and he'll probably send you home with your mother and father. This is only an arraignment today."

Somehow we got through the paperwork at the precinct, but I could not face putting the girl in a paddy wagon for the next procedural step in this cruel, unalterable process. I tried to walk casually past the desk sergeant with Janet clinging to my arm.

"Hold it, marshal," he said. "She's got to go by van."

"Yeah, right. In this case I thought I'd make the delivery myself."

"Is that in an official city car?" The sergeant had heard a million hard-luck stories. Regulations were his concern.

"Absolutely," I said. "And I'm going to get on the air to our dispatcher. It's perfectly legal."

He waved us on and I drove to the Tombs through the heavy afternoon traffic. Janet's breathing was labored. I shut out of my mind the thoughts about the next grim hours.

I had to handcuff her for the walk into the Tombs, the multistoried holding penitentiary behind the stone monster of the court.

"Don't leave me, John," she whimpered as we proceeded to the reception desk.

"More routine," I said. "They'll keep you here for a short period, then take you across to the court where I'll be waiting."

"I'll come with you." The tears were streaming from her eyes. I saw uniformed corrections officers watching us closely.

"Don't cry in front of these people." I let the words come in bursts. "Tough it out. Do what they tell you. Don't talk to anybody."

And then she was sucked into the corrections machine, brisk and impersonal; escorted by a female officer through the green barred gate and out of sight to be photographed and searched and shut in the main holding tank with assorted thieves, prostitutes, addicts, and fags masquerading as women. The fags were kept with the women for fear they would be torn apart in the adjoining male pen.

I made out the complaint papers at the court and these too went into the clogged, slow-turning mixer of due process, so that it was seven-thirty and night court before I got to the arraignment room. Janet would also be on her way there now.

Dr. and Mrs. Sakowitz were in the third row, both pale as paper. I picked them straight away and asked them to follow me out to the hall, where I related everything I had discovered about their daughter and the fires.

"Her problem," I said, "is that she cannot handle the courses at Sheridan. She was overwhelmed and with nobody to turn to because you people insisted on a medical career for her. That's why she set the fires and that's why

I'm here now arraigning a panic-stricken little girl for arson in the second degree."

I was brusque; I deserved that much release from my own pumping emotions.

Dr. Sakowitz, a tall, upright man with iron-gray hair, made no reply. He just kept shaking or nodding his head. I wondered if he knew that my intervention with them was recklessly irregular, that a good defense attorney would put in evidence that the arresting officer was of the stated opinion that the defendant was unstable, unaccountable for her actions.

"Can we see Janet?" asked Mrs. Sakowitz. She was so out of her depth in these waters that it was ludicrous.

"No, not yet. They will call her name and she'll be brought out of the side, back door. I will take her to the judge who will set bail. If you can make the bail she'll go home with you. If you cannot, she'll go back to the Tombs."

"She's in the Tombs!" Sakowitz reeled as if struck in the face.

"Yes," I said. "She's a name and a mug-shot in the assembly line of the human refuse of this city. She's in the parade with the dehumanized, if that's a word, with some who are victims of a life that has kicked them in the teeth from the beginning and some who are insane and some who are animals by preference. Look at the people in the courtroom tonight, Dr. Sakowitz, and you'll see the families of prisoners who will seem to you as much the dregs as the people who stand in the dock. Many of them are beyond caring, but others will be hurting inside as much as you are, confused and frightened, not understanding what is happening and what will happen next. Your daughter's been thrown into a zoo and my advice is to get her the hell out of here." My voice had gone husky. "Take her home and let her lead her own goddam life in her own way."

I left them abruptly and went to the bench in the front row, taking my seat with patrolmen and detectives. Whatever the columnists said, night court was hideous. We showed it in our faces. A woman police officer came over and asked along the line for John Barracato. I raised my hand.

"Your prisoner wants to see you."

Janet was in the cage behind the padded door, sitting ramrod stiff on a plank seat with five other prisoners. Her hands were clasped firmly on her lap. If anybody had touched her she would have fallen into a dead faint.

When she saw me her face glowed. One afternoon in hell and she was ready to clutch at anything familiar from the ordered, outside world.

"John, I thought you had gone away."

"No way, honey, I'm waiting for them to call your name and then I'll do my thing with the judge and you'll be out of here in no time. Your parents are in the court. They're upset but they will make the bail all right and you'll sleep in your own bed tonight."

I was not prepared for her reaction. She sprang off the seat and wrapped her arms around my shoulders and sank her face into my neck.

"I love you, John," she said.

Over her head I saw the other prisoners looking up with dead eyes. I wore my badge; obviously I was the arresting officer.

"Jesus, what a shit," spat the big black man at the end of the bench.

I gently removed Janet's arms and sat her down. I went out of the door thinking that the black man was right.

The judge was a jaded keeper of the fate, a political appointee with protectors in high places, whose only interest was in processing the flotsam that came out of the padded door as quickly as possible. My own task proved easy from the time he glanced down over his glasses at the waif before him. After I said my piece the judge remanded Janet to her parents' custody without bail.

A month later I gave evidence at the grand jury hearing. Janet seemed remote from me and showed no emotion at the indictment for arson two. I figured that if you had a wealthy physician for a father you did not have too much to fear. I did not expect to be called again for the case, and I never was.

Chapter 11

The Spangle Factory

EMPATHY between partners is as vital in a patrol car as it is in a marriage, and I found my ideal mate in the spring of 1971 when I teamed up with Eddie Duke. He came to the division a little after I did from Engine Company 209, a celebrated station that is housed in a historic brick-studded concrete building on Bedford Avenue in Bedford-Stuyvesant. We had been friends from the beginning and when Phil Winters and several other marshals retired, and there was a shuffling of partners, we were permitted to work together in Car 55, covering Brooklyn-Staten Island. In many ways we were opposite—Duke's conservatism in dress, my flashiness; his deep Catholic commitment, my religious indifference; his careful attention to and memory for the tiniest detail, my relish for the flamboyant approach. But I admired the way he operated, and he respected my style. We shared an appreciation for the ridiculous in life, laughed at the same things, enjoyed—or did not enjoy—the same people. There was a communication between us that did not need words—no explanations, no justifications. We never tired of being together and our idea of a splendid night out was to be on a working patrol. I imagined us as a couple of street cavaliers, indispensablé to the great circus of the city.

Our practice on night tours was to eat supper in one of the myriad Italian restaurants around six o'clock, for once the jobs started to come in we might not get a break until three in the morning and that could land us in some obscure diner with a previous customer's food caught in the fork and an inch-thick china coffee mug with permanent veins of dirt on the lip.

We were traveling Bay Street, in Rosebank, one night early in our association, looking for a place to eat, when we saw a cloud of thick black smoke on the western skyline.

"Think we're gonna miss supper, Eddie," I said. He was already swinging the car into a side street, aiming for the fire. Duke was a rush-stop-rush driver and we had a running gag where I would cringe down on the passenger side with my thumb in my mouth. His part was to mock my hyperactive heart.

We listened to the alarm transmissions on the radio and the battalion chief's aide telling Brooklyn dispatch that a factory had exploded and all hands were occupied knocking down multiple fires. Minutes later we pulled into a short street of old apartment houses and factory buildings. Five different pieces of fire apparatus were scattered at the far end. Hoses were stretched, and ladders extended, to a wide, hangar-like building of age-darkened red brick with about a dozen shattered windows on the front face. Wisps of the black cloud we had seen from a distance lingered high in the June sky. There was little smoke around the building itself.

Duke parked two hundred feet from the rigs and we intercepted the chief returning to his car. The factory had been closed and the employees had left work for the day, he said; no doubt about the suspicious circumstances. Several separate fires were found on the factory floor, as well as the explosion that blew the roof off. Our eyes went to the top of the building. The brick parapet looked as if it had been battered by a giant sledgehammer. Big chunks of brick and masonry had fallen to the street. The low-pitched, asphalt-coated roof, however, seemed to be in place.

"A million to one chance," said the chief. "The explosion lifted the entire roof off the joists and kited it eight feet in the air, according to the people who saw it. It fell back almost exactly in place, except that it knocked down the parapet wall."

The chief said the fires had been extinguished and the building was clear for physical examination. On the way to the front door I noticed a Ford LTD sedan on the opposite side of the street. Two bricks rested in their dents on the roof. The rest of the debris was close to the factory, so I assumed the sedan had been parked in front and moved after the explosion.

The door, well over on the right side of the building, led into a hallway that went forty feet to the back of the structure. An office suite of three rooms, fashioned out of plywood, was to our right; an arch opened to a large factory floor on the left. Thick wire lines were strung along the near wall with paper-thin sheets of silver metal folded across them; other sheets were crumpled on the floor. Much of the floor was littered with dime-sized spangles, and thousands more were strewn on wooden workbenches, along with upended metal stamps and a scatter of tin scissors and balls of thread. Over at the other wall a half dozen long, movable hanging rails were jammed with women's dresses and fancy uniforms of one sort and another, all glittering with sequins of various colors.

"What we got here," said Duke drily, "is your average, everyday spangle factory."

We paced the wooden floor and counted six different points of fire origin. At each the alligation * of the charred wood showed that the fire must have been fueled by a flammable liquid. A fifty-five gallon drum of acetone, highly flammable, was overturned by the archway, and by its side was a blackened galvanized pail—probably used to ferry the acetone. It was an extremely clumsy job of

* When wood burns the surface knobbles to the pattern of an alligator's hide, hence our term alligation. The fiercer the fire, the deeper and closer the crimp marks.

arson; nobody has pools of acetone lying about a factory floor by accident. The perpetrator obviously counted on reducing the building to a pile of ash, destroying the evidence. He had underestimated the volatility of acetone. The vapors, lighter than air, would have gone quickly to the roof and become thicker and thicker as the arsonist went about his work. As soon as he lit a match there would have been spontaneous explosions of the vapors everywhere in the factory and ignition of the liquid acetate. This explained the direct, vertical lifting of the roof and also the minor fires on the floor. Once the vapors had exploded there was not much left to support combustion. In fact, there were no extensions from one floor fire to the next.

Duke went back to the car to call in the division's photo unit for pictorial evidence of what we had found, and then we inspected the offices. Typewriters, adding machines, and calculators were in place on the desks, telling us in one glance that burglary was not a motive in the case.

One of the patrolmen from the squad car outside joined us. He was young with brown hair standing up from a pronounced widow's peak, and he was waving a notebook in the air.

"I don't know what you guys get from looking at a lot of burned wood," he said, "but I think there's something fishy going on here. Me and my partner were driving down the street when we heard the explosion, bricks flying everywhere. Seconds later we see this guy running out of the building with his pants legs on fire . . ."

"That's the man we want, officer," I said. "We think he's the one tried to burn the building down." I had been trying to figure how the person who struck the match could have escaped injury in the lightning-quick combustion of the acetone vapors.

"You do?" said the cop. "Well, no problem. We took him to Richmond Jewish Hospital. He's not going anywhere. He's the owner of the joint. Fellow named Myer Solomon. Anyway, he jumps into a car right out front and he's beating his pants legs and yowling like a wounded bear when we pull up. We see that he's badly burned and we put him into our car to take him to the hospital when we hear the fire engines. His car is blocking the front, so we—"

"So you pushed the car to the other side of the street," said Duke. He never missed anything.

"Yeah, that's right," said the cop, sounding deflated. "We push the car over, drop Solomon at the hospital, and come back here."

"Did anyone else see Solomon run out of the factory with his pants on fire?" I asked.

"Can't be sure. There were a couple of Puerto Rican girls playing handball two buildings down but otherwise the street was deserted. That was dead lucky. Nice spring night like this. Lotta people could have been out. Anyone standing under this building would have been stoned to death."

"You still have Solomon's car keys?" I cut in.

"Yeah."

"Good, let's have them. We're going to impound it. The car was used in the commission of a crime—as the intended getaway vehicle."

Duke and I went over to the Ford LTD and searched it thoroughly. On the floor of the driver's side we found singed particles of the man's pants and these we dropped into plastic sleeves for use as evidence to corroborate the officer's story. In the trunk was the real evidence of premediated arson, a large manila envelope containing typewritten lists of the factory inventory, dated the day of the fire, plus estimated costs. The total amount was $235,406. From our observations some insurance company was going to be hit with a wildly exaggerated evaluation of a loss caused by the client himself.

"Lousy sonofabitch," I said to Duke. "One thing I can't abide is a fraud fire. A guy gets mad at his wife and sets the couch on fire. That I can understand. I don't like it, but I can understand it. But anyone who makes a fire like this just for greed, just for money, I lose my patience. Never mind the people who live upstairs or next door and stand a good chance of getting burned to death, and never mind the firemen who have to risk their lives to fight the fire. The bastard only cares about his money. They're the lowest type of human. You'll see. This Solomon will come on like a saint, a Bible in one hand and a picture of his crippled old mother in the other."

"Cool off, John." Duke had a soft, intent way of speaking that could charm the thorns out of a cactus. "When we get to the hospital, let me do the talking."

Solomon was still in the emergency ward, sitting up in a screened-off bed with his legs swathed in bandages and a yarmulke on his head.

"See," I said to Duke through gritted teeth. "The beanie cap. Was he wearing that when he splashed the acetone around?"

We approached the bed on the same side and showed our shields. Solomon had a bony face with long flat cheeks and soft purple lips. I guessed his age at fifty-two.

"Mr. Solomon," I said. "How are you doing?"

"Ohvey, ohvey." He didn't say anything else and this peeved me as much as the hypocrisy of the yarmulke.

"Mr. Solomon," I went on, "Let me put it bluntly. You fucked up."

His brown eyes flared at me. "What do you mean?"

"I mean you tried to burn your place down and you did more damage to your miserable body than you did to the factory."

"What talk is this?" he said, puckering up.

I felt Duke's hand on my sleeve. "All right, Eddie, you talk to him."

"Mr. Solomon, we'd just like to know what happened," said Duke in his gentlest voice.

"Ohvey, who knows what happened? At five o'clock I am closing up and driving my foreman to his car. Back I come then to the factory to make out the payroll checks."

"Very good, Mr. Solomon," Duke encouraged. "You made out the payroll. We didn't know that."

"Yes, and I was in my office writing out the checks and I smelled this funny smell. So I get up and investigate where is this funny smell. I look in the ladies' room and in the men's room and then I cross over into the factory. Ohvey, all over the factory is fire. I am stamping out the flames and the explosion comes. Boom! I am knocked back by the explosion and then I see my pants have caught on fire and I am running out of the building . . ."

"Let me get it straight," said Duke. "First came the fires and after that the explosion?"

"Yes. I was in the factory stamping the flames." Duke shrugged at me. The sequence was impossible.

"How was business, Mr. Solomon?" I could not keep out of it.

"How could business be?"

"Right," I said sourly. "How could business be making nothing but goddam spangles."

"This guy I don't like," said Solomon, giving Duke a hurt look.

"Was the factory well insured?" Duke sounded sympathetic.

"I don't know. My wife takes care of all the books." He ran the tip of his tongue around his lips. "Sir," he said to Duke, "could you tell me. My car. How is it?"

"The car's fine, Mr. Solomon."

"Good. My wife is going to pick it up on the way to the hospital."

"No she's not," I said. "The car is impounded." I motioned Duke out of the ward. We borrowed an office to call the district attorney's office and convinced the prosecutor on duty that we could nail Solomon in court on the evidence of his own story, the police witness, and the inventory in the trunk of the car. He authorized us to make the arrest.

"You take the honors, John," Duke said as we walked back to emergency.

The spangle-maker was slumped back on the pillows, his yarmulke tilted over one ear.

"Mr. Solomon," I said. "You are under arrest. The charge is arson in the first degree. You have the right to remain silent. You have the right to consult an attorney . . ." I went through the whole Miranda warning while he nodded acknowledgment and whimpered like a victim of cruel fate.

The next day Duke and I found the two Puerto Rican girls who had been playing handball. They had seen Solomon with his pants on fire, trying to escape. Their testimony, added to the rest of the package, got Solomon indicted by a grand jury in record time, but the defense lawyer kept stalling the actual trial for months afterwards. The delay stretched to a year, then eighteen months. The girls, meanwhile, from transient families, had disappeared somewhere in the tenement forests of the city. The young patrolman had left the force and was attending law school out in Illinois. We were well aware that the defense strategy was to wait as long as possible in the hope that the witnesses would scatter, thus making it a tougher case to prosecute and increasing the chances of copping a plea. That was the way it worked out. Solomon ended up with a year's sentence for arson four. He did, however, lose what he prized most—money. The East Coast Insurance Company, which had recently reinsured the spangle factory, did not have to pay out a penny.

Every time Duke and I talked about this case I became so worked up that we had to change the subject. The legal system under which a man like Solomon could so readily sidestep true justice was screwy enough, but the factory explosion was also dressed up as a cute episode for some New Yorkers to chortle about. The day after the fire a newspaper columnist confided that the job was done by Morris the Torch. Morris had told him personally. I wondered why newspaper guys like that never moved their butt into the real world; the greatest city in the world to report and they stayed in a dark corner spinning idiot tales.

My attitude to fraudulent arsonists, as I explained to Duke, had been indelibly colored by Dr. Simon Collison, a super-slick dentist I had encountered early in my marshal days. I built an iron-mesh case around Collison but he still managed to slither through. The only satisfaction I had was that he was obliged to move out of New York State and hung up his shingle in Connecticut.

My first meeting with Collison was at seven-thirty one morning on the sidewalk outside a two-story professional office block in the Fort Hamilton section of Brooklyn. Smoke was pouring from the second-floor windows and six prostrate firemen were being administered oxygen.

Collison, who had that second arrived on the scene, was clutching his forehead and saying, "What happened? For God's sake, those are my rooms."

I introduced myself and told him I would make an inspection as soon as the smoke cleared. He was to remain outside until I came back. I did not want to add any more. The battalion chief had called in the fire as suspicious because a rear door to the block had been burst open and a substance like varnish had been spilled about the dental rooms upstairs.

The lock catch was indeed twisted away from the jamb at the back door as if struck or kicked from the outside. Upstairs were two suites of offices, one vacant, the second comprising a sparsely furnished waiting room that led to a short corridor onto which three separate rooms opened— one an office, the others each containing dental chairs. At the end of the corridor was a toilet and a file room containing steel lockers.

Fire had been set on the dental chairs, the desk, the typewriter table, and in several places on the floor. Obviously somebody had poured pools of flammable liquid in each place and touched them with a match, either as an act of nuisance sabotage against the dentist or in the expectation that the fires would come together and reduce the entire floor to rubble. The liquid used must have been the clear liquid varnish Marvelon because there were two empty cans with this label in the waiting room. This varnish gave off volatile fumes which caused heavy acrid smoke on combustion. This accounted for the debilitated firemen. As it turned out, the separate fires flashed and then burned themselves down to mere smoldering. The actual damage was slight. The dental chairs were salvageable, although to my wary patient's eye they were outmoded and probably needed replacement anyway. The office was spartan and I was surprised to find it devoid of medical and dental books. This observation led me to examine the suite again. For a dental office it was ill-equipped and had a decidedly impoverished appearance. Nowhere, for instance, could I find an x-ray machine. Nor did I locate any of the framed graduation diplomas professional men usually have in their rooms.

Back with Collison on the sidewalk, I went through the routine preliminary questions. He had had late patients, he said, and did not close up the previous night until ten o'clock. From the office he had gone to spend the night at his girlfriend's house and heard of the fire when his wife called early that morning. She had been telephoned by people who lived opposite the office block. My eyebrows raised at the man's domestic arrangements. He explained that he and his wife were legally separated. A smooth man, this Collison—the face and voice of a pitchman for an up-market television commercial, immaculately dressed in an English-cut gray suit with a fine red stripe, age about thirty-eight.

His shoes and the cuffs of his trousers, I noticed, were splattered with a hard, shiny substance. His son, he said, had sprayed them with a new shoe polish and had done a sloppy job. It was a silly lie because he had just told me he was living with his girlfriend. And who had one's shoes sprayed when they were on the feet?

Assuming that his rooms had been demolished by the fire, and with me acting the thick-headed cop, Collison went on to bemoan the loss of dental chairs worth thousands, a medical library, and a new x-ray machine.

"Gee, doc, I'm afraid somebody was out to cause you harm," I said. "In my opinion the fire was intentional. Have you been having problems with anyone?"

"To be frank," he said, "my partner and I were at odds about the practice and we dissolved a few weeks ago. I doubt very much, though, that he would do such a thing."

What a bastard, I thought. Puts his partner in a pit and throws him a smile and a length of string.

"Okay, doc." I abandoned the fumbling nonsense. "Why did you have all that Marvelon upstairs?"

"Oh yes, after my final patient last night I decided to varnish the floor to give it a good fresh shine."

"But the stuff was also painted or poured on the chairs and tables?"

Collison sucked his lips and blinked his eyes. "All right, the truth. I did varnish the floor. When I finished I sat back to admire my work and lit a cigarette. Suddenly the whole place was on fire. I was very frightened. I ran downstairs, hopped into my car and raced over to my girlfriend's house. Nobody called me the rest of the night so I figured the fire was out—that is, until my wife got on the phone this morning."

Collison was incredible; one idiot story after another. When we moved to the precinct, and I demanded his pants for laboratory tests, he tried giving us the facts. His partner had owned most of the good equipment and had most of the regular patients. After the dissolution the practice fell off and Collison was strapped for cash. He broke into his own office building in the early hours of the morning to make it appear like a burglary and torched his rooms to collect the heavy insurance assessment that had applied before the split.

In my naiveté of those days I assumed the case would zip through the courts and Collison would have his avarice cured in prison. What he did, however, was to sign a waiver of immunity at the grand jury hearing and testify that he had signed the confession of arson to protect his fourteen-year-old son. His son had stopped by the office that day for money and, receiving a rebuff, had gone away mad. Collison had thought therefore that it had been his son who set the fires. Such paternal nobility won him acquittal, and never mind the inconsistencies.

Months later I bumped into Collison in a clothing store, where he was eyeing a two-hundred-dollar mohair suit.

"Hi, doc," I greeted him. "How's things?"

"Pretty good," he said, fingering the mohair and giving me a friendly smile. "You know, I never did get back my license to practice in New York, but I'm licensed in Connecticut and I'm in the process of moving up there."

"Well, well, God bless Dr. Collison," I said. And then he went his way and I went mine.

Chapter 12

Fire and the
Naked Lady

EDDIE Duke and I had both learned from experience that the legal maze could protect the guilty as well as the innocent. We had worked long hours to seal in Myer Solomon with pictures, affidavits, scientific test evidence, inventory examinations, and book audits. And, in concert with this, we were juggling several other difficult cases. Advanced fatigue overtook us at two o'clock one morning and we went back to the bunk room. It had been transferred from the Municipal Building to the second floor of the venerable, single-engine firehouse on Duane Street in lower Manhattan.

Ninety minutes into our sleep the telephone rang. I half opened one eye. Let it not be Brooklyn, I prayed.

Joe Wilensky, the sandy-haired marshal in the nearest of the cots, grunted and threw back his blue blanket. He swung his legs over the side and sat there for a moment in his jockey shorts, letting his eyes focus in the faint light. Front and left of him, between the row of steel lockers and grime-streaked windows that looked out on Duane Street, was a scatter of ten cots and cheap ladder-back chairs draped with clothing and gun belts. Five of the cots were humped with the forms of men, probably semi-awake like me, lying still and listening.

Wilensky padded through the opening in the line of lockers in his bare feet and lifted the telephone out of the cradle.

"Yeah, fire marshals." He listened a while. "Hold on, I'll get them." He put the receiver on the table under the hooded desk light and headed back to his cot.

"Car 55." He made the announcement lightly, pleased that the call had not been from the Bronx. Wilensky covered the Bronx for this tour with Eric Figueras.

I came out of my cot and walked back to the telephone. The luminous hands on my watch showed 3:40.

"This is Barracato. Give me the particulars."

"We got a fire for you," said the voice that never slept. "Coney Island, occupied multiple dwelling."

"Why is the chief making it suspicious?" My arm was tired just holding up the telephone.

"Set fire. Third floor hallway."

"Okay. We're responding. ETA thirty minutes."

Duke was pulling on his pants when I went back to my cot.

"Coney Island, Eddie. No details, but it sounds small. Could be a grudge."

I dressed quickly and followed Duke's lean, boyish figure to the wooden stairs behind the lockers.

On the ground floor we skirted the engine, gleaming red and silver and filling up almost the entire space, and went through the hatchcover in the main swing door to the street. The three Plymouth interceptors at the curb might fetch a thousand dollars between them at a used car lot, I thought. We shared a match for cigarettes and got into our car. Duke drove through the narrow, darkened streets and out under the arcing lacework of the Brooklyn Bridge to the Gowanus.

"Wilensky told me tonight that Tom O'Neill was quitting," said Duke. "Got himself a job as a security chief at one of those big Miami Beach hotels."

I saw the signs to Staten Island and worried whether I could ever keep up mortgage payments on the new house.

"How would you like it, John?" Duke was saying.

"Oh, yeah, Miami Beach. No, not for me, Eddie. I gotta have New York City."

"Me too," said Duke.

We got off the expressway at Coney Island Avenue and made a right on Neptune.

Up ahead a pair of headlights suddenly veered across our side of the road. Duke banged his foot hard on the brake pedal, making the tires squeal. He was a champion with the brake. The headlights in front swung to the right and we saw the unmistakable contours of a Cadillac. The Caddy tried a U-turn, but the lock was too wide, and it smashed into a line of brimming garbage cans on the opposite sidewalk. A hubcap was knocked loose from a back wheel and went spinning across the pavement to clatter against the iron grating at the entrance to a delicatessen. The driver of the Cadillac, a heavy-set man in a light tan jacket, got out and started weaving around the back of the car to retrieve the hubcap. His foot missed the curb and he went sprawling on the ground among milk cartons and egg shells and apple cores. A red-headed woman in a short skirt stepped out on the passenger side.

"For crissakes, Marty," she complained.

Duke rolled our car closer. "Hey, pal, are you hurt?"

"S'okay, s'orright," said the man in the light tan jacket.

"Eddie," I said, "the guy's bombed."

Duke got out of the patrol car, walked to the Caddy, and took the keys out of the ignition. He unlocked the trunk, tossed in the keys, and slammed the lid down.

"Get a cab home," he said to the drunk. "You're likely to cause somebody harm."

The man peered at Duke from a sitting position. "Thanks, ossifer. Li'l much to drink."

"Yeah, thanks," said the woman. She had been studying me from a distance. "That's a good suggestion."

I gave Duke a pleased look when he got back into the

car. "Cute, Eddie, very cute. Charlie Brewer would have upended that drunk in one of those garbage cans and rolled him down the street. I don't hold with the rough stuff."

Duke put the car in drive and accelerated down Neptune. "I never understood it either," he said. "We do a hell of a lot better our way."

The place was a weary four-story brick apartment house of pre-1900 vintage, four apartments on each floor. The fire rigs were gone but lights were on in many of the windows. A bull-necked man in a tartan hunting shirt, who met us at the front door and identified himself as the superintendent, showed us a pile of water-soaked fire debris against the wall in the hallway by 3D. The newspapers and cardboard and plastic wrappers, in various stages of disintegration, would have thrown a great deal of smoke. The damage, however, was minimal. The lower half of the walls was tiled and the floors marbled. I picked it as the work of a pyromaniac who lived in the building, probably on this floor. It was the wrong location for a hired torch job or a revenge fire. I did not speculate aloud. Duke's thoughts would be going the same way.

"Folks in the building are pretty upset," said the superintendent. "Here it's after four in the morning and they're scared to go back to bed. This is the third fire on this floor. One burned itself out and I killed the other with pails of water. I called the Fire Department tonight because you couldn't see this hallway for smoke."

"You had problems with any of the tenants on this floor?" Duke asked him.

"Jesus, where do you want me to start." He ran his hand through uncombed black locks. "The apartments rent for ninety-five a month so we don't attract too many Supreme Court judges. In 3D, for instance, we got two lesbians, Jackie and Jill, for crissakes, they call themselves. The guy in 3A is a recluse who keeps roaches for company. Open his door, you'll faint. I swear he must shit on the floor and piss in the bath. But we can do worse than that. The woman in 3C is a screaming alcoholic, and when I say screaming, I mean she gets a load on and hollers at her boyfriend like she's gonna cut his balls off. In fact, that's what she says. You can hear her a block away screaming that he's a fairy 'cause he never lays her. In the mornings she's all right. Sick with a hangover, I suppose. She even acts shy. At night she goes off like a rocket. The boyfriend is a hound for the bottle, too. He doesn't sleep here all the time, comes and goes, you know. He told me once that Rita, that's the screamer, was beaten half to death by her father when she was a kid because he caught her screwing in the back of a car. She won't look anyone in the face when she's sober. Soon as she gets into the gin she's stopping people on the stairs and swinging her hips. She's propositioned every man in the place, me included, and someone told me she put the hard word on Jill once."

"You ought to go into psychology," I told him. "You've left out 3B."

"Empty. A young couple moved in on a Monday and moved out on the Friday. Said they couldn't live next door to a sewer. What can I do? We had the health inspectors up here. They go in holding their noses and come out holding their wallets. The guy probably has a million dollars stashed under his mattress."

Since the building seemed to be awake anyway, Duke and I went ahead knocking on doors and asking questions. The superintendent was right about the recluse, an unwashed man about forty years old in a stinking apartment that made my flesh crawl. He said he saw nothing, knew nothing, cared nothing. The lesbians were very young, perhaps nineteen, both students at New York University. Their three and a half rooms were clean and tastefully furnished. We noted they had a double divan bed, covered with a white fur throw rug, against the east wall of the living room. Jackie did the talking while Jill, a peach-skinned blonde, sat cross-legged on the floor stroking a black and white cat.

Jackie told us that she thought Rita was the firebug. She

had not seen her actually setting a fire, but Rita was crazy and capable of anything.

"I've looked through the peephole late at night and seen her naked in the hall," she said. "She's drunk, of course, and she's usually shouting at that man she lives with."

"What does she shout?" Duke was a stickler for detail.

"Last time," said Jackie, "she was ordering him to get out of her fucking home." She pronounced the "g" carefully.

Rita Panizza—Jackie had given us the last name—flung open the door after one knock. She was unsteady on her feet and her short dark hair straggled over her ears. I guessed she was about thirty-five going on fifty. She seemed to be wearing only a white cotton housecoat. It was damp and the nipples of her breasts poked at it.

"What yuh want?" she demanded.

While we told her we were there to investigate the fire, she swayed in the doorway, sending blasts of sweet gin fumes into our faces. She looked Duke up and down, and then me.

"Hey, you're some fancy guy," she said. Her voice had dropped an octave. "You don't belong here. Building's full of creeps. Everybody hates everybody in this house, did yuh know that?"

"What about the fires, Miss Panizza?"

She leaned over so that her mouth was close to my ear. I wondered if you could get drunk on gin smell alone. "The lesbos," she whispered. "I hear them through the wall. Jeez, do they go at it. I think they light the fires to cool off."

She giggled and put her arm around my shoulder. I gently untangled myself. "We'll find the fire-setter," I told her. "You can bet on it."

For the next couple of hours we questioned other tenants in the building and they came up as a surprisingly normal group of people. Mrs. Ferris, a Nordic-looking divorced woman with three children who lived in 4A, re-

vived us with breakfast. She asked if we could do anything about the recluse who lived directly underneath.
Roaches from his apartment were scrambling through the
floor boards into hers. We had to tell her we were confined to fire investigation. The woman in 4B, Joan Petersen, joined us for coffee. She was Mrs. Ferris's cousin and
lived alone. Duke asked both women enough questions to
write their biographies. They seemed not to have any
sinister shadows in their past to suggest pyromania.

We left our telephone number with every tenant and
promised to return if there were more fires. By conserva-
tive estimate we had ten calls to the building in the next
month for small garbage fires in the hallways of the third,
and now the fourth floors. Also, they were being set in
front of doorways and not in the open hall spaces as before. Jackie and Jill had a fire at the door and so did the
recluse and Mrs. Ferris. None of them spread because the
doors in this building were metal-sheathed.

Rita was starting to enjoy our visits. "Hi, John, Eddie,"
she greeted us after the tenth fire. "Come in and have a
belt."

"Forget the booze," growled Duke. "We know positively that you're setting the fires, Rita. Why don't you
give us the full story?"

"Why me, Eddie?" She thrust out her breasts when she
was into the bottle, and she never wore a brassiere.

Because you're the only one around here who hasn't
had a fire at the door."

It was such an obvious trap that we did not think Rita
would fall for it. We overestimated her. Three nights later
the superintendent had us back examining a fire at Rita's
door. Partially burned copies of the *Daily News* had been
swept to one side for our inspection.

"Now they've got it in for me," said Rita. "I smelled
smoke around about nine tonight when I was in the
kitchen and when I went looking there were flames coming up from under the door.

Scorch marks on the door proved that the fire had been

inside as well as outside. We asked Rita to stay put while we went into the hall and closed the door. A weather-stripping lip of metal was screwed to the bottom. I tried to push a plastic credit card between the lip and the floor. It would not go through. Perhaps, with infinite patience, someone could have threaded newspapers under the door, one sheet at a time, but it was much more likely that the newspapers were spread over the sill, lit, and the door forced shut on them.

We knocked on the door for Rita to open up again. "You lit this fire yourself," Duke accused. "We couldn't get a credit card under the door, let alone a wad of newspapers. Own up to it, come on, we're not going to hurt you. We've discussed it and we agree you need hospital care. You're falling apart, Rita, and we want to help you."

Rita's sallow face mottled in rage. "You bastards!" she shouted, and hurled the door in our faces.

"That's it, Eddie," I said. "This kind of circumstantial evidence would never hold up in court. We'll have to stake her out."

Joan Petersen on the fourth floor agreed to move in with her cousin and let us use her apartment for a mid-night to 6 A.M. watch on Thursday, Friday, and Saturday night of the following week. The fires were invariably set on one of these three nights. Duke took the first three hours on the Thursday. He sat on the stairs leading to the roof, a hidden position from which you could hear every sound in the third and flourth floor hallways. Meanwhile, I took a nap on Joan's couch, my head on one arm rest, my feet dangling over the other. I dreamed I was getting a heart attack, the vise closing tighter and tigher on my chest. I awakened in a heavy sweat and gazed directly into a pair of green eyes with bright yellow bars at their center. A big white angora cat sat on my chest, grinning at me. I pushed her away, feeling stupid, and joined Duke on the steps. "Can't seem to doze off, Eddie. Think I'll sit it out with you."

At 2 A.M. we heard the rustling of paper somewhere

below. Our heads came up; our eyes locked. I felt an irrepressible urge to laugh. Two fire marshals sitting on a stoop in a sleeping building, startled like fawns at the sound of rattling paper. Duke's eyes were dancing too, and his cheeks puffed out with suppressed laughter. We both had to climb to the roof to let it all out, like schoolgirls caught peeping at a grownup party.

We never did know what caused the noise. There were no fires that night, nor on the next two nights, nor on the following Thursday. We told Joan to move back into her apartment; this could take weeks and we were comfortable enough on the stairs. The Friday was Flo's birthday and Duke and his wife Shirley took us to dinner at the Amber, at 80th and Third Avenue in Brooklyn. It's a cozy Italian restaurant inside a friendly bar in a mostly Irish neighborhood. We gorged ourselves on the excellent veal and shrimp scampi, consumed two quarts of table burgundy, and finished with demitasse coffee spiked with anisette. I drove Flo home, returned for Duke, who had taken Shirley back to their place in Bay Ridge, and we staggered to our fourth-floor roof staircase in Coney Island on the stroke of twelve. I sat lengthwise across the second step with my back to the railing. Duke took the top step. We both fell into a dead sleep and spent the six hours oblivious to everything. Somebody could have burned the building down around our ears. Fortunately, they did not, and when we crept downstairs at six-thirty we saw with relief that there were no fresh scorch marks in the halls.

Vincent Canty telephoned me at home Saturday to call off the stake. It was summer, the division was short-staffed, and there were too many other pressing cases.

"You have yourself in another Rockaway situation," he said.

"By Christ, so I have, chief." The words shot out of my mouth. In a flash I remembered the long hunt for Terence Kelly and the ultraviolet. We could use the same thing on Rita. Even if we were not on the scene, the invisible col-

ors should stay on her hands and feet for a few hours until
we did get a summons.

"Calm down, John." Canty sounded stern. He hated
blasphemy.

"Sorry, sir. We'll give it one more night and that's it."

On the way upstairs at midnight we sprayed Rita's
doorknob and the threshhold with blue ultra-violet, shone
the lantern to ensure it had taken, and took up our posi-
tions.

We heard raised voices and the crash of breaking dishes
from Rita's apartment at two o'clock. The boyfriend must
be home, we thought. Everything quieted down in the
building for twenty minutes, then we heard a door open
and close and the rasp of slippered feet coming up the
stairs. We peered over the railing of our hideaway. Rita
came into view, stark white naked except for fuzzy pink
slippers. She carried two brown supermarket bags overflowing with shredded newspaper. One she placed outside Mrs. Ferris's door and the other outside the door of
4C, which was occupied by a frail, elderly woman we had
rarely seen.

I put a restraining hand on Eddie. We had to see fire.

Rita struck a match to one bag, crossed quickly and
fired the other, then was galloping down the stairs two at
a time.

We hit the hallway at the same time Joan Petersen's
door opened. She must have heard Rita's feet thumping
on the stairs.

"Oh, my God!" she shrieked. "Fire! Fire!"

"Get some water, Joan," I shouted.

She disappeared back inside her door. Mrs. Ferris
popped out of hers, causing the burning bag to flop into
her hallway. Simultaneously, the elderly woman opened
her door. She saw the fire at her feet and swooned against
the door post.

Duke lunged for her while I stamped on the fire. Joan
was dousing the blaze at her cousin's door with a pan of
water. Duke put the old lady in Mrs. Ferris's arms and

grabbed the ultraviolet lantern from the roof steps. Blue footprints came and went from the two apartments and down the stairs. We followed them to the door of 3C.

"Open up, Rita, we got you this time," I hollered.

"Piss off," yelled Rita.

"Open the goddam door," commanded a male voice from inside.

"Open it yourself, faggot," snarled Rita.

The boyfriend, clad in pajamas, finally swung the door wide.

Duke marched to the center of the room and faced Rita, who had wrapped her scrawny body in a housecoat. "You are under arrest. The charge is arson in the second degree. You have the right to remain silent . . ."

"Yeah, yeah, yeah," said Rita. She stared over Duke's shoulder. The fourth-floor tenants, including the old lady, were assembled in the doorway watching.

"Hold out your right hand, Rita," I said.

Puzzled, she obeyed. The lantern showed the blue on her palm.

It was close to six when Duke and I had Rita alone in a glassed-in office off the precinct squadroom. We had an airtight case, but we preferred a frank admission. In any event we wanted to know more of the woman's background to recommend psychiatric evaluation before the court. She was completely sobered and talking quietly. Her eyes would hold Duke's or mine for an instant and then slide off, her chin rubbing against her shoulder. Her childhood had been a living terror because her stepfather, a heavy-drinking longshoreman, beat her with his closed fists for the slightest misdemeanor. One night the stepfather caught her in the back of a car with a boy. He pulled her out of the car by the hair and rammed her head into a fence post. As she lay dazed on the ground he kicked her so hard in the buttocks that the wound ulcerated and she had required surgery. She had run away from home later and worked as a waitress, living with a series of men. Alcohol became a crutch for her. As the years went by, and her looks and figure went with them, men lost interest. Her latest boyfriend would rather drink whisky than sleep with her. She freely told us how she built up Dutch courage with the gin and solicited other men in the building. They pushed her aside like a dirty rag. Often at night she would lie awake listening to the sounds of the lesbians at their lovemaking, and this would get her so stimulated that she took more gin to deaden the desire.

"Either that, or get your satisfaction from a fire," Duke said gently.

She shook her head. "I didn't set those fires."

I excused myself and went to a delicatessen around the corner for black coffee and bagels. When I came back Duke had his hand resting on Rita's. I put two coffee containers and the bagels on the table by their side and left the room.

From the squad room I saw the tears start falling down Rita's cheeks. Duke got up and put his arm around her, and I knew that she had at last admitted to the fires and trusted us to do what was right by her.

At the arraignment we persuaded the judge to dispatch Rita Panizza to Kings County Hospital for psychiatric interviews. Two examining physicians sent her back to the women's detention barracks, ruling her fit for trial. Her lawyer had the charge broken to arson four and Rita returned to the building in Coney Island on five years' probation.

For months after that we had periodic calls from Joan Petersen, who said she could smell smoke. One night we made a detour to talk with Rita. Her boyfriend had moved out permanently. Her apartment stank of gin and wet, burned papers.

"Rita, you're still at it, aren't you?" Duke put on his disappointed paternal frown.

"Honest, Eddie, I'm being good. You guys picked the perfect time to visit. I'm about to take a shower. What say we all hop in together?"

We laughed and she laughed with us. Her face crinkled into a hundred lines. She seemed to have aged years.

"She needs a thorough drying out and six months in a

psychiatric hospital," I said on the way out. "Those god-dam shrinks who checked her in the summer must have been as spaced out as she is. I'll wager she's setting fires in the bathtub or the kitchen sink, getting off in some peculiar way, and then turning on the faucet."

"We can't intervene unless she takes to the halls again," said Duke.

A full year later I was telephoned at home by Cornelius Stanley, one of the marshals who had covered the night beat in Brooklyn. A small rubbish fire set on a third-floor hallroom had caught some packing crates standing nearby and the top floors of the building were gutted. All residents were rescued by aerial ladder. A half-dozen tenants had accused a Rita Panizza, who occupied 3C, and said Barracato and Duke had the history on her.

"This Rita Panizza was having a fit of hysterics at the scene, John, and we had her psychoed to Kings County," explained Stanley. "But the shrinks said she was rational and told us to take her away. That's where we stand. We don't have enough for an arrest."

"What was the name of the examining psychiatrist, Corny?"

"Dr. Marcus Lesseur."

"Okay, you keep Rita at the hospital, wait five minutes, then take her back to Lesseur."

I hung up, dialed Kings County, and got connected to Dr. Lesseur.

When I told him who I was, I heard him make a clicking sound of impatience.

"Doctor, you have just seen a patient named Rita Pa-nizza. She has a long record of alcoholism and pathological firesetting and we want her admitted for the full evaluation and treatment. She needs to be cured. You understand that, doctor. Now if you send her away again and she sets another fire, and people die in that fire, we will see that you never practice medicine again in New York as long as you live."

"I don't take kindly to threats, marshal," said Lesseur.

"And the city doesn't take kindly to psychos running loose in the streets burning people out of their homes."

"All right," he said. "We'll take another look at the woman. Send her back."

"She didn't leave, doctor. She'll be in your office in one minute."

Rita Panizza was duly admitted to the psychiatric ward and we heard she was transferred to a mental hospital upstate. Too little is known about the disease of pyromania and I doubt that it can be totally cured, but at least, belatedly, Rita was being given a chance to break from the cruel prison of her former life.

Duke and I caught so many psycho cases that we joked about winding up in the asylum ourselves—a gag that became reality in the winter of 1972.

The director of the Indian Creek School for the mentally retarded in Bensonhurst, Brooklyn, personally called on our squad superintendent, Jim Susman, about a rash of fires that had occurred over recent weeks in the office and maintenance buildings of his institution. The director said the prime suspect was a former inmate, Bobby Higgins, aged nineteen, who had been released to society some months earlier after spending most of his life at Indian Creek. Psychiatrists had decided that his intelligence had improved to the point where he no longer needed confinement. However—and the director was defensive about this—Higgins had no family on the outside, no skills, no place to go, and only a subsistence welfare allotment. He was continually sneaking back into the asylum for food and shelter. Frequently he would climb into the dormitories through windows, shift an inmate from bed to floor, and use the bed for the rest of the night. Higgins was a big lad—six feet, two inches, two hundred pounds. On some occasions he hid in the corner of the cafeteria and commanded others to bring him food. Higgins' raids were tolerated until he began setting fires for reasons, the director supposed, of pique and larceny. Twice he was seen running from small fires in the office block where cash drawers had been rifled. A secretary's desk jar, containing loose change, had been stolen the second time. Security guards had often given chase, but he would outdistance his pursuers and disappear without trace into the woods at the perimeter of the Indian Creek grounds.

The director did not want to call in a police manhunt for Higgins because the institution was under the gun from one of the New York television news programs for shoddy housing and inadequate care for the patients. A local newsman was leading a crusade to "clean up Indian Creek" and was arriving at all hours of the day and night with lights and cameras. If the Higgins story got out before the crusading newsman made his name and went to Hollywood, it would be distorted and sensationalized. The director conceded that Indian Creek was halfway to a snake pit, but outsiders had no notion of the deranged incontinence of his charges and the difficulty of getting dedicated staff. The place was designated a school. In fact, the seven hundred persons institutionalized ranged from small children to men and women in their sixties. Cadging a favor as an old Army buddy, the director asked Jim Susman if he could assign a couple of intelligent, fleet-footed marshals to the task of bringing Higgins to heel, confirming the arson, and putting him away for psychiatric re-evaluation.

"Barracato and Duke, get in here," Susman called from his cubicle office. And that was the start of our asylum caper.

One thing about Susman, he did not tell you how to do your job. Duke and I talked over the assignment at length and Duke thought that it would be best to stake out the place in the guise of inmates without telling Susman or the director in case it was forbidden or they insisted on protecting our welfare by informing the Indian Creek security guards. If Higgins was as elusive as the director claimed, he was also cunning. We would not get near him as cops or as obvious undercover men. Brownsville had taught me that you go all the way on a stake, and Duke had scored one of his notable successes while acting the role of a counterman in a bodega in East Harlem.

We cased the place the next day by car, posing as visitors. Indian Creek consisted of seven numbered brick three-story buildings, close together, set in about twenty acres of lawns and woods surrounded by a high, barb-topped cyclone fence. On the east side, most distant from the buildings, a thicket of maples and gums nudged the fence. This was Bobby Higgins' escape route, and little wonder. The cyclone link lifted up from the ground as easily as a skirt. We decided to park our car late each afternoon on the eastern public road, go under the fence and through the woods, and stroll up the service road to the main clutch of buildings.

In the winter time the adult patients at Indian Creek wore dark brown woolen military coats of World War I vintage and we found a couple of these at a disposal store, along with army folding caps that were also in vogue at the institution. Thus attired, we emerged from the woods and shuffled up the service road on Tuesday night to begin the hunt. We made a pact not to speak to each other unless absolutely necessary and under no circumstances to laugh.

The night was weird but uneventful. We wandered around Building Five, where the older inmates were housed, then went to the cafeteria in Building Three to stand in line and receive free dollops of mashed potatoes and beef ragout. It took an hour to pass the serving section. The people would be handed a plate of food; then they would spend forever putting bread rolls on and off the tray, emptying the salt shakers, and pushing backwards and forwards for spoons and forks. Duke and I sat at a table by ourselves and mumbled over our food, our appetites beyond recall. Bobby Higgins did not show. We had memorized his picture—a wide-shouldered black youth with hair like a thin fur covering on his head, thick and prominent lips, and heavy forehead.

On the following evening the recreation hall was set up for the regular midweek dance and we put in an appearance in case Higgins elected himself some amusement.

We were deafened and struck dumb the moment we walked in. Fifty or more of the inmates were twisting and turning and stamping in ecstatic tantrum to the fast beat of the rock music coming from a stereo placed at center stage. The wail of the guitars had them throwing their arms in the air like worshipers at a Southern revival meeting, the throb of the drums set them beating a tattoo on the floor with their heavy boots. Others, however, merely dragged along, singly or in pairs, making an occasional skip and then pulling back to a knock-kneed foot shuffle. And there were dazed souls too, standing about in furtive, round-shouldered huddles, not daring to abandon themselves to the wild music. It was pitiful and spooky. Misshapen girls in their late teens drooled from the mouth in excitement, mongoloid men from sixteen to sixty rolled their eyes and cackled in glee.

"I can't take this, Eddie," I muttered. We were standing just inside the door. "Let's get the hell out and look for Bobby tomorrow."

Duke moved away instead of replying. I saw the reason. Bearing down on me was a young man with his head thrust way out of his collar and his tongue between his lips, and he was guiding a mongoloid girl of about his own age. She wore a wrinkled green cotton dress and walked pigeon-toed.

"Hey," said the young man putting his face against mine. "You want to dance?" He stepped back and pushed the girl between us. "She wants to dance with you."

I felt the blood drain from my face. "Him," I croaked, pointing at Duke. "He wants to dance."

The man tugged the girl over to Duke. "She wants to dance," he repeated.

Duke drooped his shoulders and went slack-jawed. "I don't dance," he whined.

He stayed there shifting from one foot to the other as part of a threesome watching the grotesque floor show.

Slowly then, he backed up to my side. "Sonofabitch. What was I supposed to say to her?"

"It wouldn't matter what you said. They don't understand anything."

We loitered there for ten minutes longer. I was losing my revulsion as I studied the dancers. The expressions on some faces was positively seraphic. The young man who had approached with his mongoloid companion, I noticed, was holding her hand.

"You know, Eddie," I said, "no matter who you are and what you are there's always somebody for somebody. And that's a pretty nice thing."

We left the dance and prowled around the other buildings until eleven. On our way out down the service road we were spotted in the headlights of a touring security car. It came abreast and one of the Indian Creek guards got out on the passenger side.

"Hey, what the hell are you guys doing out here this time of night?" he asked in the mildly cross tone of a man whose job it is to round up straying children.

We gave him our best imbecile look. "Okay," he said. "Which building do you live in?"

"Five," I replied.

The guard put us in the back of the car and his partner drove us to Building Five. We were taken through the double glass doors to a large steel table in the center of the lobby where a buxom, white-suited female black orderly sat among a profusion of files and dispensary bottles of pills and various colored liquids.

"I found these two aimlessly wandering around," the guard told the orderly.

"Tut, tut," said she. "Naughty boys. Take your medicine and go straight to bed."

She pulled two little plastic cups from a box at her side and poured into each about a jigger of purple fluid.

"Tank-ou," said Duke, mimicking a kid with a cold. He took his cup and downed the contents in one gulp. I had no choice but to do likewise.

We dawdled off down the corridor, turned the corner,

and plunged into a men's room. It was fifteen minutes before the lady watchdog finally left her post and we were able to scoot through the front door and cross-country to the woods. We were using my car. I got Duke home all right, but I had to fight all the way to Staten Island against the lead weights on my eyelids. The last part of the journey did not register on my consciousness. The next I knew Flo was shaking me vigorously by the shoulder.

"John, wake up!" My eyes opened on the alarm clock. It was eleven-thirty the next day.

Late that afternoon we went through the cafeteria line and took our trays to a corner table. There, in the diagonally opposite corner table, wolfing a plate of beans, was Bobby Higgins. Duke circled from one side, me from the other. Our stealthy approach must have warned Higgins that we were not harmless inmates, for his eyes narrowed on me and suddenly he was off, weaving through the tables and out the door. In fast pursuit, we saw him dash behind Building Four and duck under the pylon-raised rear section. He cowered there, showing us the whites of his eyes.

"It's okay, Bobby, we're here to help," I crooned as I crawled in after him with my handcuffs. "Just slip these on, baby-san." The expression I had picked up in Korea came out involuntarily.

Higgins accepted the cuffs on his wrists without a word. He was docile, pleased even to surrender his fate into other hands.

A small crowd had followed us from the cafeteria and when security guards arrived to investigate we identified ourselves. They recognized Higgins and were glad to let us take the full responsibility.

On the way to the precinct I went over the allegations against him in my mind and concentrated on the occasion when he had apparently entered the office block at night, set fire to papers on the top of a desk, and stolen, among other things, the secretary's change jar. She had said it contained about twenty dollars.

"Bobby," I said, as we turned up Kings Highway, "you should be ashamed of yourself for setting those fires."

"I don't know about no fires, man," he replied. His response was no different than I would have expected from any street kid. Higgins was borderline deficient, an I.Q. of seventy, I understood. "I was just eating," Higgins went on. "I got to eat there 'cause I got no money."

"Yeah, well that was the most stupid thing you did, Bobby. You try to burn down a whole office to steal a jar containing a few pennies. That ain't gonna buy you much, Bobby, a few pennies."

"Bullshit to you, mister," said Higgins in an offended voice. I had tapped his pride. "There were nickels and dimes and quarters in that jar too. I got over twenty-three dollars outa there."

We had the lever then to coax the rest of Higgins' story of his arson and burglaries at Indian Creek.

It was slow in coming and we did not force the pace. We wanted a careful assembly of facts to recommend psychiatric testing. At one point in the questioning in a precinct anteroom Higgins maintained that he had seen two other inmates set the fire in the office. He had pulled the alarm. His dossier showed, however, that his fingerprints had been found on the window sill and on the glass door to an inner office.

"Bobby, you must have been there," said Duke in his low, soothing voice. "Bobby? Bobby?"

I was at the water cooler with my back turned when I heard Duke's voice rising. I turned to see Higgins' big, woolly head nestled on his chest. He was sound asleep.

A detective lieutenant loomed in the doorway and studied our sleeping prisoner. "I knew you guys were low-key," he said, "but isn't this taking it a bit far?"

Higgins went straight from court to Kings County Hospital and the last we heard he was in a mental institution near Kingston, New York. The crusading reporter got his own television show, and Duke and I let Jim Susman's assumption stand that we just happened to walk into Indian

Creek one night and came face to face with our prey. If there were rules about outsiders parading around an asylum disguised as inmates, we did not want to know about them.

Chapter 13

Undercover

IN the early summer of 1973 a plague of burglaries hit the bars of Staten Island's South Beach, a seaside stretch of washed-out pastel cottages, sweaty amusement parks, and concession stands. It would have been exclusively a police matter, except that the thief had a penchant for setting fires at his departure from each place. Sometimes he would put a match to a pile of papers in the office, on other occasions he would toss a Molotov back through the window or up onto the roof. It was his exit line.

My amazement never ceased at the variety of ways humankind could get itself twisted up. But the bar owners did not want their South Beach Spoiler psychoanalyzed, they wanted him stopped. The peak season was approaching and they needed the uninterrupted business. Duke and I were given a free hand by the police and fire departments to find the fire nut and we happily and literally dodged underground.

At the outset we concentrated on Ackerman's Inn, a purely drinking establishment with a high, triangular front and a mixed clientele of blue-collar workers and some fringe characters who operated as loan sharks, hijackers, and drug runners. Someone had busted into the bar through a side window and emptied a hidden drawer under the cash register of seventeen checks that had been

cashed by the bartender through the course of a long, liquid evening. The burglar had also stacked a couple of
chairs against the bar and set fire to them with newspaper
kindling. The proprietor of a check-cashing place down
the street had subsequently admitted to paying fifty cents
on the dollar for the checks. He did not know they were
stolen, he insisted, and readily consented to identifying
the passer from the mug shots. Unfortunately, the extensive gallery of police pictures failed to include the guilty
party, so we decided to stake out Ackerman's with a long-
range camera and get prints of all the regulars. We figured
the thief was familiar with the bar because he had known
about the hidden drawer. Of course, he could have given
the checks to a confederate to cash, but we had to start
somewhere.

The division has a green Dodge surveillance van with
side windows that are transparent only if you are on the
inside looking out, and we parked this opposite the bar
for the photography. To explain the day-long presence of
the vehicle we borrowed Con Edison railing fences and
traffic cones to screen off the manhole in front of the van.

On the first morning of our set-up I manned the long-
lens camera while Duke clambered out of the truck in a
Con Edison helmet and tool-laden belt to open the manhole cover. He carried lengths of cable and wire which
we planned to throw in and out of the hole to give the
illusion of repairing electrical conduits.

Five minutes after Duke went to work he was tapping
at the front window.

"What's up?" I hissed at him.

"We goofed," he hissed back. "The manhole doesn't
lead to the electrical conduits. It's a goddam sewer hole."

"Who's gonna know the difference?"

"Me, for one. It stinks like you wouldn't believe and
you can see the rats' eyes glowing in the dark."

"We got no choice," I said. "It's the only manhole within
range. Get back in there and fake it."

Duke put in an hour in the hole, play-acting with his
pliers and cables, and then insisted on swapping jobs.

The sewer odor was clinging to his clothes. I could smell it through the open window. "You'll foul up the van," I protested.

"You won't notice if you stink too," he said.

For the best part of three days we alternated in the truck and the hole, looking forward to the rush hours when the one on sewer duty could transfer to directing traffic around our bottleneck. We snapped 350 pictures of Ackerman's clients. They did not turn up our quarry, but they did cause consternation in the Police Department. One of the photographs showed an undercover narcotics detective; others were of men he had personally arrested months before. Miraculously, they had not come face to face. The narco cop was pulled out and reassigned.

We abandoned our sewer to investigate a fresh burglary and fire in the Beer Mug bar and grill, a long, low, white brick veneer place on Seaside Boulevard. Up until midnight the bar was overflowing with residents and visitors, coming and going under the spotlights that picked out wooden replicas of foaming beer mugs spaced around the roof. By one, however, the place was deserted but for a few drunks sprawled against the side of the building. The burglar had broken through the back door shortly after one, rifled a safe that a patient child could open and, in his fashion, had thrown a Molotov on the roof of the bar as he fled. A neighbor had pulled the alarm minutes later and the fire was quickly extinguished by the men of Engine Company 161 and Ladder Company 81—my own old outfit. My brother Leon still worked with the engine company. I had a strong personal interest in the fire at the Beer Mug.

The bartender told us that it might be fruitful to question Tom White, a local whisky-head who had run up small tabs at a number of bars in the area. White had been refused further credit at the Beer Mug the previous night and had expressed his displeasure to the barman before leaving.

"He didn't get nasty, though," Duke suggested.

"Tom never gets nasty drunk," said the barkeep. "He's

one of those guys who gets super-polite when he has a load on. That's the thing; you don't really know what's going on in his head."

Around noon we found White in a bar a few blocks west, sipping a tall Scotch and water. He was neatly dressed, about fifty-eight years old, with a red, pitted face and gray hair.

"I'll cooperate with you gentlemen in any way I can," he said after we showed him our tins and explained our mission.

"Were you in the Beer Mug last night?"

"No, sir."

"Okay, Tom," I sighed. "Down to the precinct."

After two hours of prompting we confirmed that White had been asked to leave the Beer Mug, had gone to the rear of the building for a leak, and had gone home. He had not broken in, and he was frightened of fire. We turned him loose.

Two nights later the same bar suffered another burglary and fire, this one in the restaurant area. A shotgun with an elaborately carved stock, used by the owner Arnold Keough for duck-hunting, was stolen from the office. Once again Tom White had been in earlier, unsuccessfully cadging drinks.

When a bar search failed to flush him out, we asked around the stores and eventually found White in a funeral parlor, kneeling before the open coffin of an elderly woman who had evidently been a close former friend. With due reverence, we waited until he had finished his silent mourning. He startled us when he got up and came toward us. His hair was combed, his shoes shined, his black pinstripe suit freshly pressed and cleaned. At our first meeting he had been neat; now he was glossy enough to be the undertaker himself.

"Remember us, Tom?" said Duke. Old Tom had no idea who we were.

There was only a faint odor of whisky about him when we returned to the precinct, but years of steady soaking must have drowned many of the key cells in White's brain. He could not recall the last time he had visited the Beer Mug, though he was prepared to accept our word for his movements. In fact, he was agreeable to any of our suggestions short of arson and burglary. Duke thought the vagueness might be an act; I put it to the test.

"Okay, Tom, we've had it with the bullshitting," I barked, rolling a fingerprint card into the typewriter. "We'll type up the arrest card and you can tell your story to a judge. It's gonna go harder on you this way, but you give us no choice."

"Gosh, I'm sorry," he said. "it's just that I don't know anything."

White gave his full name, date of birth, place of birth, every detail without protest. I pretended to type them up. What I punched out instead was: "Eddie—what the hell are we going to do with this s.o.b.? We got zero on him."

"Here, Eddie," I said, whipping the card out of the typewriter. "You gotta witness this."

Duke read the note with a grave face and excused himself from the squad room. He was back in five minutes, resting a hand on White's shoulder. "You're lucky, Tom. Our boss thinks you deserve another chance. You're free to go."

For our next gambit we moved on two fronts. First, we instructed the bartender at the Beer Mug to drop the word that Keough was offering a big reward for the shotgun because it was a family heirloom. We fixed a tape recorder to the bar telephone and told him to flick it on if anyone called about the gun. Our suspicious minds had not ruled out the possibility that Keough had taken advantage of the crime wave and "stolen" the gun himself to collect insurance. The gun would be hard to fence because of the unique stock carvings. There was every chance the thief would try to sell it back to the owner if there was a thief.

The second play was a stake-out of the Beer Mug, which had been hit three times in the space of a month. I volunteered to pose as a drunken bum, sprawled in the doorway of the bar from midnight to dawn. Duke would position himself with filter-lens binoculars on the board-walk some two hundred yards away. The bar was on a main boulevard corner. There was a clear view across open parking lots and a bus turnaround area. If I signaled he was to come running. We decided that it was essential for one of us to be on the premises to eyeball the arsonist in the act and to stop him from escaping in the pitch dark-ness of the thickly reeded sand dunes in back of the tav-ern.

I made a good wino; unshaven, dirty sneakers, dungarees, ragged teeshirt. Spilled Thunderbird gave me the appropriate stench. For good measure I doused some on the front of my fly as if I had pissed in my pants. I did not wear a gun, a badge, or a transmitter, just a dollar in my hip pocket in case I was mugged by a junkie. Coming up empty-handed, he might put a knife in my ribs out of pique.

On the fourth or fifth night I was tugged out of the doorway by my heels by members of the South Beach street gang. Duke said his binoculars fogged up he got so excited. The bastard did not think of galloping to the res-cue, however. He watched while the young hoods rolled me over on the sidewalk, removed my dollar and kicked my ass. Duke was not going to blow a perfectly good cover, he explained later, merely to save a dollar.

The next incident gave him a bigger charge. The gang pulled a false alarm on the box in front of the Beer Mug. Engine Company 161 responded and the firemen prowled around the building, my brother among them. I lay with my face in the shadows of the doorway and an arm crooked across my eyes like a little child who figures that you can't see him, if he can't see you.

"Must have been some goddam kid or maybe that bum in the doorway." I recognized the voice of George Graf-ton, the chauffeur. "Some filthy skel. He's even pissed himself."

Duke worried that he was getting all the fun on this

stake. One night, about 2 A.M., he signaled me with his flashlight. I made for our car, hidden under a tree in the parking lot, as prearranged. Duke came strolling along with a broad grin on his face.

"A gesture of appreciation for the greatest comedy act in South Beach," he said, opening up the back door and producing a white styrofoam ice chest. He spread out a feast of burgundy, cheese, and crackers, and we spent an hour toasting ourselves for service beyond the call of duty. Back in the doorway I saw that the distant, beautiful lights of the Verrazano had doubled in number. I blew sweet burgundy breath in their direction and slept like a baby.

The reward offer for the shotgun paid off first. A caller said a "certain party" had the gun and would discuss a deal with Keough the following Wednesday night by the maintenance shed on the boardwalk at Arrochar, a mile or so up the line. When Keough heard the tape he bet ten dollars to a quarter that it was his former cook, Lonnie Scaglione.

Next day the *Advance* carried a picture and story of a man who had been arrested—and released on bail—trying to hold up a liquor store with a toy gun. I was incredulous; the man was identified as Orlando (Lonnie) Scaglione, 28, a cook, of 423 Shetland Drive, Arrochar.

Duke and I drove over to his house and waited two hours until he came out and began walking up the street.

"Hold it there, Lonnie," Duke called, coming up fast behind and showing his badge in his hand. "We got some questions for you about Arnold Keough's shotgun."

Scaglione spun around and glared from under the fleshy hoods over his eyes. He was an ugly sonofagun. "You fucking guys never leave a man alone, do you," he spat. "Now it's a shotgun. I don't know shit about any shotgun. Pick on someone else, you rotten bastards."

"Good old, Lonnie," said Duke. "Got a million of them, haven't you, pal. Let's take a ride."

Down at the precinct I played the tape of the telephone

call. "You want to eat the tape, Lonnie, or do you want to cooperate? We'll book you for the burglary and the arson both."

"Oh, no, you can't get me on the arson. I didn't set no fire. I got the shotgun off of somebody else."

"We want it all, Lonnie."

"Okay, okay, okay. The straight story is that a guy by the name of Frank Labriotta broke into Keough's joint, stole the shotgun, and set the fire. Frank sold the gun to me for fifty dollars and then I heard from the bartender that Arnie Keough was willing to pay three hundred to get it back. So I made a few bucks, what's the big deal?"

We took Scaglione to his house to get the shotgun.

"You can't come in, fellas," he said. "My mother's a fine old Italian lady and she'd get upset."

"Got a million of them, ain't you, Lonnie," Duke drawled again. "All right by us, but carry out that gun by the tip of the barrel, just like you're holding a snake by the tail."

Scaglione delivered the shotgun as ordered and we proceeded to lose interest in him.

"You're not going to believe this, Eddie," I said when we were driving off, "but Frankie Labriotta is an old friend. Friend in quotes. I arrested him about five years ago for setting fires in a pizzeria and a Singer sewing machine place in the South Beach Plaza. He busted into these places, picked up anything of value, and set a fire. I didn't connect him with this lot of bar fires because I haven't heard of him since. Matter of fact, I thought his folks had straightened him out. His father owns Labriotta's Fuel Oil Company. When I picked him up his mother tore strips off me for being an Italian guy making trouble for a nice Italian boy. He was sixteen years old then, a strong kid with a wide face and tiny black eyes that made you wonder if he could see properly out of them. Anyway, I got him to court and asked the judge to send him off for psychiatric evaluation. I can't recall whether he got it or not, but I do remember Mrs. Labriotta showing up in court for the preliminary hearing and accusing me of persecuting her son to the extent that the kid had run away. He wasn't in court because the parents said they did not know where in hell he had gone. Mrs. Labriotta said I was the one who should be on trial because I was strictly Gestapo. And that, Eddie-san, was the end of the saga of Frankie Labriotta until this very day."

A couple of discreet telephone calls gave us the information that young Labriotta was driving an oil truck for his father's company, and we interrupted him with handcuffs while he was filling the basement tank at a crumbling old mansion house in St. George. Same Frankie; eyes like black buttons on a broad olive face.

On the way to the station house I asked Labriotta if he remembered me booking him for the two fires back in 1968. He said he did.

"I was always curious," I said. "Where did you run away to?"

"My aunt's over in Jersey."

"Did that make sense, Frankie? We could have stopped all this fire stuff when you were still a youngster. You would have had professional help to get you on the right track. Now here we are five years later and you're gonna be photographed, fingerprinted, charged in the courts with burglary and arson, and probably sent to jail. Don't you think we could have prevented this, Frankie?"

"Yeah, maybe you're right."

"Okay, as long as you know."

Jesus, I thought, what a straw I was clutching at—to get through to this young man's mind that he could have been turned away from criminal impulses, that he had a sickness requiring treatment. Doubtless he realized that with his old man's money and political clout he would not have to spend an hour behind bars if he burned the Staten Island ferry with a hundred people aboard.

After the session at the precinct, Duke drove me home. "That's the last we'll hear of Frankie's case," I said. "His father's too heavily hung in this borough."

We had a lot of fun on the caper, and no regrets. We understood the city. What the hell, the bar fires ended and South Beach had a good season. Nobody ever makes a perfect score in New York.

Outsiders who looked at the horrendous crime picture in New York, and the grievous inadequacies and injustices of the correctional system, were puzzled at the irrepressible zeal of guys like Duke and me. Where's the satisfaction? they asked. Your impact is negligible, so why bother? One answer was a lovely Puerto Rican lady of twenty-nine named Irene Ajeron.

Irene ran a small bodega on a corner of New Lots Avenue in Brownsville, and that same summer she called us in to chase off an extortionist who intended bleeding her dry. Irene's husband had been shot dead in an armed robbery at the store two years earlier. Although her family had pleaded with her to return to Puerto Rico with her two daughters, aged four and six, Irene had determined to keep the business going. She had counter help from high school students, but she did the lion's share of the work. From seven in the morning until nine at night she served in the store, did the buying, and ran backwards and forwards from her apartment upstairs to tend the kids. Her grit did not go unrewarded; numerous families in the blocks around trimmed their supermarket lists to give Irene a share of their business.

She was clearing two hundred dollars a week and everything was swell until the mysterious night fires started. The first one was a small gasoline blaze at the front entrance grating, the second a similar piece of arson at the side delivery door.

The day after the second fire Irene had a visitor, one Reggie Troop, a young thug who had been in and out of jail for rape and attempted murder. Troop said he knew who was setting the fires and could have them stopped if Irene paid him $150 a week protection money. Irene told him to get lost, and that night a Molotov was dropped into a rear storage cellar. Hundreds of dollars worth of groceries were destroyed.

When Irene related the full story to us in the store next morning, we suggested she arrange a meeting with Troop and pay him off. We would provide $150 in bills marked with ultraviolet and we would wire her for sound to get Troop on tape, making his pitch.

Even as I was laying out the plot I had second thoughts. She looked like a teenager to me standing there behind the counter in blue jeans and red checked shirt, her hair tumbling about her smooth brown face. There was a hardness to her eyes, the whites like marble and the color like enamel, but you had to study her to see it and I sensed that they had not always been this way.

"You understand, Irene, that you'll be running a risk," I said. "Troop is an animal, capable of anything. You don't have to go through with it. If you do, the meeting will have to be in an open place. The alternative is for us to stake out the store every night and catch Troop in the act. It's certain that he's setting the fires himself."

"What is the best way to get him?" she asked.

"For us to conceal a recorder and sending unit under your clothing and for you to make a meeting. We'll be close by and we'll jump on him the instant he takes the money."

"I trust you," she said. "I shall do it."

We collected the necessary equipment from Manhattan and telephoned Irene to put the store in care of one of her high school casuals and meet us at the old firehouse around the corner on Maple. Troop might be watching the bodega. He would anticipate one visit there by fire marshals, but two would get him agitated.

In an upstairs room of the firehouse Duke and I hemmed and hawed around Irene for a long time before getting to the task. We had not wired a woman before.

With a man it was easy; whip off the shirt and tape the recorder and transmitter to his bare belly with a big Ace bandage. The recorder wire extends up the front and the tiny microphone is taped to the chest. The twelve-inch antenna for the sending unit is usually taped up the backbone.

Irene pulled her shirt out of her jeans and Duke attempted to reach in underneath with the Ace belly bandage. She laughed at his fumbling and stripped down to her bra.

"Do it right," she said.

With fidgety fingers Duke fixed the bandage and tucked in the recorder and transmitter, ran the antenna up her bare back, and, with a shrug of resignation, fed the recorder wire under the bra and taped the mike in the cleavage between her breasts. Her figure, to say the least, was glorious, and Duke was red-faced when he finished. Except for his wife Shirley, with whom he was madly in love, I don't think Duke had been this intimate with a woman before in his entire life.

While Irene returned to the store we called in three day cars from the division to position themselves at different points four blocks distant. Duke and I waited in the surveillance van two hundred yards away on New Lots and set our frequency to the mike at Irene's bosom.

Her voice came over the receiver as clear as a local telephone call. We listened to her making small talk with customers for the next two hours. Around midafternoon Duke shot forward and killed the box for a minute. We had heard the unmistakable sounds of Irene going to the toilet. At four-thirty a female voice had us tuning up the maximum volume.

"I be's Reggie's woman," said the voice. "You got sumthin' youse want me to tell him?"

"You be's a no-good bitch," we heard Irene reply. The visitor started to say something, but Irene cut her off. "I got the money. Tell him that."

It was dusk before the same female voice came back on the wire. "Reggie be's going around the block. If'n he sees the man then he gonna cut your ass. If'n he don't he waits at the church, corner of Hickory."

"Ain't gonna be no man," said Irene. "I'll meet him."

My palms were sweating when I checked my gun. "Hickory's four blocks east," I said to Duke. "That makes it a quarter mile. We'll start rolling when we hear Reggie."

We would be flying blind. Our view of the front of the bodega was obstructed by two parked trucks and we could not identify people on the sidewalk from that distance in the dark. Duke radioed the other three patrols that the payoff was set for nine o'clock in front of St. Teresa's. They were not to move in until we gave the signal. Everyone had studied Reggie Troop's mug shot. He was a massive black man with a flat, broken nose, and was never without a buckskin vest. This meant, to us, that he carried a knife in his back belt band.

Traffic noises buzzed over Irene's microphone and we knew she was out on the street.

"I'm going to meet him now." Irene spoke in a whisper. We imagined her tucking her head down to talk directly into the mike. She was some Puerto Rican lady, this one.

My heart set up its ba-boom, ba-boom. The receiver played us the roar of buses and unintelligible street chatter.

Then it was Irene's voice. "I'm here." It wasn't whispered this time. She had met the mark.

"Against the wall." The reply was a deep bass. It had to be Troop.

Duke turned on the ignition and we pulled slowly into the light stream of traffic.

"Why?" Irene sounded fearful now.

"I don't want to be seen," said Troop.

There was a pause. Duke was pacing us carefully. We were on the same side of the street as the church.

"If I give you the money will you stop making the fires?" Irene was superb. She knew we needed the admission.

"Yeah. You got the stuff?"

"Will you stop the fires?"

"No more fires, you know that. You just keep paying me, chick, and I don't make no fires."

Duke gunned the motor for the last block. I barked over the radio: "Close in! Close in!"

I saw Troop then by the church door lights, an enormous figure next to Irene, pushing a wad of bills into his breast pocket. My feet hit the pavement the instant Duke braked and I cake-walked across the sidewalk, my gun extended and held high. "Freeze, you sonofabitch!"

Irene peeled off to one side as Troop's hand went to his back.

He saw Duke coming around the front of the van and let the hand fall.

"Hey, man, you got the wrong guy," he said. His eyes were going from Duke to me. His brain was working. "I ain't done nothing. I jes' talking to this chick. Ain't that right, honey. Tell the man. We be's old friends."

We heard the sirens coming closer and closer.

Troop's feet were shuffling toward Irene and his hand was going back to the knife.

"Cut the jiving, man. We'll blow your head off." Troop was a killer, but I had yet to shoot a man. If he made the move, I didn't know if I could waste him or not. A crowd was gathering to the right and left. I started going forward, Duke at my side.

Suddenly, Troop smiled and put his huge hands out where we could see them. The lights and sirens of our three back-up cars were all around us, mounted up onto the sidewalk.

"Suckers," he said. "You got nothin'."

"We got this," I said, reaching into his shirt pocket and retrieving the money. "One-fifty in tens, and all of them marked."

Duke pulled Troop's arms behind him and wrestled to get the cuffs on. Even on the last link they cut into the beef of the man's wrists.

I walked Irene back to the store, where she stripped off her bandages and wires.

"Beautiful job," I said. "We heard everything you did, everything you said."

A light peach flush came to her cheeks. "Everything?"

I gave her an exaggerated leer. "Everything."

"You one bad cop," she said. She punched me in the chest, but she was smiling.

Irene was on the telephone to us next morning and this time she sounded furious. "You better get down here," she said. "It's not finished."

We raced back to Brownsville with our sirens blaring. Irene had a wide-eyed black man of about thirty sitting on a kitchen chair by her counter and she was prowling up and down in front of him like a guard dog.

"This one," she fumed, "is telling me that it was Leo down the block that hired that bad-ass Reggie to burn me out."

"Leo's Deli?" I asked.

"You got it," she said. "My opposition, except that he's got a liquor license and he's into check cashing."

I put my attention on the man in the chair. "What's your name, man?"

"Ferdie Newcombe."

"Okay, Ferdie, what's the scoop?"

"I heard you got Reggie Troop and I can help you fix him more than one way. Sonofabitch tried to burn me out night before last. I know what it is, man, to look out the door and see flames coming up the steps at you. This guy is a sickie, yes? I'm nobody's angel but I can't understand no man raping a woman and I can't understand a man wanting to burn up another human being and his family, somebody's who's totally and completely innocent."

"Reggie's done both," I said, "and we're going to get him."

"You listen, man. Like Reggie owed me money and I approached him. I says, 'I want my money, Reggie,' and he says 'I don't have any money.' I says, 'Oh, yeah, you means you burned down half the fucking neighborhood and you're broke? What you doing this for, fun?' He goes, 'Leo paid me money to do a job, but I didn't do it thorough, so I gives the money back.' I tell him I'm gonna get my money one way or the other, outa his ass or outa his pocket. Then he takes me to Leo's and him and Leo go to the back of the store and down the staircase into the cellar. Leo says for me to stay up front, but there's a soda machine by the stairway and I go over there. I hear Reggie say, 'I need money to get this guy off my ass.' And Leo says, 'You don't get no more money until you finish the job. We arranged a fee for you to burn up Irene and you didn't do it right.' "

I interrupted, "You heard him say that?"

"Sure. I'm not sitting here telling you this for my health. Okay, then I hear Reggie say, 'What about the money from the checks?' Leo tells him he's got to wait 'till the checks clear. I've had enough of this and I call down to them, 'Reggie, you better get my fucking money.' Leo jumps at this and says, 'I told you to wait at the front of the store.' I tell them I'm not interested in their conversation. I just want my money. When Reggie comes up he says he'll pay me when the checks clear. I give him one more day and that night he sets a fire in the basement of my goddam building. That's why I want to stick him and Leo both. One sonofabitch is greedy and the other one is sick enough to let himself be used."

Duke came in with the main question. "You be willing to tell the grand jury what you told us here today, Ferdie?"

"Yeah, I'll come."

"You ever been arrested, Ferdie? Any heavy stuff?"

"No, no, no, no."

Ferdie Newcombe did appear before the grand jury and we got indictments and later convictions that put Troop away for another ten years. Leo got one to five. It was some den of thieves and cutthroats. Our star witness, Newcombe, it turned out, was a convicted murderer who stole welfare checks and gave them to Troop for cashing in Leo's store at one-quarter value. Newcombe's sole interest was revenge on Troop for copping the bread.

Irene's there still, doing nicely, and the lights are back in her eyes since she remarried. For as long as Duke and I were together we used to stop by to make corny jokes about checking her wiring. She joined the mischief and said that we could do a lot worse. We have no argument about that.

Chapter 14

The Young Savages

EARLY one evening in late August two thirteen-year-old black kids sauntered along the curbside of Marcy Avenue, in Bedford-Stuyvesant, peeking through the windows of parked cars to see if anyone had been fool enough to leave a key in the ignition lock. They wore denim jackets with the likeness of a smoking revolver brush-painted in white on the back. A number of young girls on Marcy recognized them as Tonto and Bones, two of the junior members of The Pistols street gang, whose turf ranged across about six blocks of Marcy and Tompkins on the north side of Atlantic Avenue. Tonto was a small lad with a copper hue to his skin and straight, jet black hair; the grandson, allegedly, of a Cherokee Indian who had once worked as a steeplejack on the Empire State Building construction. Bones was six feet tall, feet like plates, and weighed 120 pounds. Local legend had it that in heisting a store the gang would slide Bones under the door to slip the bolts.

This night they were going freelance, looking for a car to strip. It was a mistake because they were not yet sufficiently experienced to know which cars were untouchable. Obvious pimpmobiles and the Saturday night dudemobiles of the drug pushers were given a miss, but the run-of-the-mill working vehicles of the district's main man pushers were unfamiliar to them. They were not aware that the gray 1967 Oldsmobile parked in front of Morgan's Shoe Store was the distributor car for Joey Lark, a heroin trafficker with a line direct to the higher echelons of the mob. The stuff came to Morgan's in the hollow heels of new shoes, it was later established, and Joey Lark made regular pick-ups.

Bones spotted the keys in the Oldsmobile and jumped in behind the wheel while Tonto scrambled into the passenger seat. They snaked through back streets to a vacant lot on Junius, in Brownsville, and commenced removing the battery and radio. When they went into the trunk for the jack to hoist up the side wheels, they found a brown paper bag filled with glassine envelopes of heroin. The discovery scared the hell out of them. They were trying to figure a way to get the stuff back to the owner when the situation was taken out of their hands by two big black guys who had been watching from a stoop across the street.

"What you got dere, man?" The two dudes had come up unseen in the darkness.

"A sack of groceries is all," muttered Bones.

The two men took the sack to the streetlight for close inspection. "They be's right," said the spokesman. "Ain't nothing but groceries."

He pulled out his wallet and handed Bones a twenty-dollar bill. "You boys could get into trouble round here at night. You take off. We'll look after these groceries."

Bones was shaking so much he was about to rattle. The diminutive Tonto was speechless with fright. "Git," said the man. His companion swept down and plucked a blade out of his boot. The boys didn't stop running for five blocks.

Next morning Joey Lark showed up at the basketball court at the side of Lincoln High School. He wore a maroon velvet suit and he had a fire in his eyes that could melt metal.

A half dozen youths were having a workout on the

court, sweating and panting in the August heat. Their
jackets were hanging on the cyclone fence, each with a
picture of a smoking pistol on the back.

"Mungo," Lark called icily. "Get your ass over here."

"Yessir, Mr. Lark, yessir." Mungo was the leader of The
Pistols, a stocky, eighteen-year-old man with the raised
welt of a knife scar down his right cheek. Main men of
Joey Lark's stature and wealth were Mungo's idols. He
Uncle Tommed them shamelessly. With lower life he was
a vicious punk. It was a commentary on his wretched subculture values that he threw on his jacket before he
dashed over to Lark. As gang boss Mungo was entitled to
wear a fur collar; he wanted to remind his distinguished
visitor that he was one of the coming men in the neighborhood.

"I'm gonna tell you something once, Mungo, no more,
just one time," said Lark. "Last night two of your sonofa-
bitching Pistols stole my car on Marcy Avenue. That was
a bad thing to do, Mungo, and what made it worse was
that I had stuff in the back—a lot of stuff. It is now Tuesday. If I don't get the car and stuff back by Friday, then
you and every other member of your gang is gonna get
dimes on their eyes.* You know, Mungo, that there is no
way this won't happen. You got me?"

"Jee-sus, Mr. Lark. They be's crazy. I'll handle it personal. You good as got everything back . . ."

Mungo lapsed into silence. Joey Lark was already striding off down the sidewalk, his tall, elegant frame being
followed by the respectful eyes on the basketball court.

Within the hour the gang seniors had rounded up all
members and beaten a confession out of Bones and Tonto,
whose terror had been obvious from the moment they had
been called to the schoolyard. They took a second beating
when they said the car was certainly stripped to the
chassis by now and the heroin had been taken by
strangers.

Mungo left Bones and Tonto bloody and bruised at the

* be killed

school and moved over to Jefferson Park with six others for a strategy session.

"We gotta waste them," said Mungo. "Ain't no way Joey Lark gonna accept anything less. I don't wanna be no dead man. We take a vote. Beat their heads in with baseball bats or we burn them in their beds."

Mungo asked each youth in turn; the decision was to burn.

The gang members spent the rest of the day making Molotov cocktails and storing them in a basement clubhouse. They used pint whisky bottles and gas drained from the tank of a truck. Mungo had the job of stuffing in the wicks.

Shortly after 1 A.M. they rendezvoused at the clubhouse, collected the Molotovs, and walked over to the three-story brick-and-joist tenement where Bones lived on Tompkins. The boy's real name was Henry, the oldest son of Lucille Jamieson, a welfare mother with three other children. The Jamieson family shared a five-room apartment on the third floor; mother and two daughters in one bedroom, Henry and his kid brother in another, with the grandmother sleeping on a living room hide-a-bed. The children's father had deserted three years earlier. Mrs. Jamieson was frequently heard scolding Henry for the company he kept and for his late night excursions. She was a good mama and never stopped cooking him minor feasts to build up his skeleton body.

The only other occupied apartment in the building housed the Cox family on the second floor—the mother, Mavis Cox; daughter Susan, 18 and pregnant; daughter Cloris, 16; and three sons, William, 22, Carter, 20, and Randolph, 12. The Cox household was a noisy, crowded dormitory of pleasant young people who had been raised on welfare. The older ones were now setting themselves up with good jobs, and street gossip had it that three hundred a week was going into a family bank account towards a home over on the Island.

The rooms on the ground floor were empty and strewn

with rubbish. Two side windows, shoulder-level, were broken, and the lock on the front door snapped open with a hefty push. If ever there was an easy physical mark for arsonists the tenement at 418A Tompkins was it. On the human level, however, the place seemed too bursting with life to be torched without warning, except perhaps by a lunatic or a practiced Nazi exterminator. The Jamiesons and the Coxes had played a lot of years in a very tough game of survival, but they were making out, and the young were still at the threshold of life.

Yet the Pistols, young black lads from similar families, set about their job of retribution against Henry Jamieson as if the tenement was an empty shed. One of the members, Lemon Tree Murray, a boy of fourteen with naturally puckered lips, was posted as lookout on the opposite sidewalk while Mungo and his lieutenants went into the hall and ground-floor vacant rooms with their Molotovs. Each took a different position and, at a low whistle from Mungo, the boy in the back room fired a wick and dashed his bottle to the floor. He retreated to a former bedroom at the left, where a second gangman threw his Molotov. A third member exploded his gas-filled whisky bottle in the parlor, and then they fled out into the hall and past Mungo, who smashed his Moltov on the stairway. Outside, Mungo raced to the side of the building, and, at the signal, two other youths tossed Molotovs through the broken windows.

The ground floor was furiously ablaze and the boys were gone into the shadows when Samuel Walker, who lived across the street, saw the fire from his front window and called in the alarm.

Mavis Cox, awakened by the crackling sound of flames and the smell of smoke, shouted to her brood. They crowded to the fire escape window in the front but the way was blocked by flames curling up from the windows below. William Cox herded everyone to his back bedroom, tore the mattress from his bed, and stuffed it out the window. The mattress fell in the courtyard, hard against the building wall. Carter went first, showing how to drop straight down to the soft bed of packed fibre. He landed on his feet and jumped nimbly aside, beckoning and calling to his mother to come next. Flames roared at the side of the tenement and smoke was gushing from the back entrance. Mavis Cox paused for a moment on the sill of her son's room, then dropped in a flailing of arms and legs. She landed flat on her back, her lower spine catching the side of the mattress. Carter picked her up in his arms. She was unconscious, her back broken. Susan hit the mattress on her belly, her unborn child killed instantly. The others made the leap as neatly as Carter. The Cox family was out alive.

What happened on the third floor could only be conjectured. Somebody, probably Lucille Jamieson, heard the fire and swung open the metal door to the hall. By then the blast of heat and smoke was murderous. The door might have held it back for another fifteen minutes if it was closed. In the panic it was left ajar, letting the monster inside the apartment.

People in the street saw grandmother Jamieson at the front window, crying to them. "Wait! Wait!" they called. The sirens were screaming in the night. Another minute the ladders would be up. It was seconds later, though, when the grandmother fell out of sight, overcome by the carbon monoxide. Lucille Jamieson aroused her two daughters and took them to Henry's room at the rear of the apartment. She put her four children on the bed and lay across their bodies, protecting them, gallantly, hopelessly, from the fire consuming her house . . .

At 2:19 A.M., Duke and I were cruising along Pitkin Avenue, listening to the early cryptic radio reports on the fire at 418A Tompkins Avenue. It was a warm summer night, a guaranteed dozen fires in our area, and we had made a drowsy wish that none would require our urgent attention.

"Do you think the city would notice if we hit the cot for a few hours?" suggested Duke.

"Yes, but we'll do it anyway," I said.

Duke wheeled around onto Rockaway, headed for Atlantic Avenue and Manhattan. The radio barked a rebuke.

"We have people trapped in the building," the aide to the 29 Battalion chief told Brooklyn dispatch. His voice croaked. He had breathed too much smoke.

Duke flipped the siren without a word. Simultaneously I had the bubble light on the roof. A surge of adrenalin cured our lassitude.

"We're attempting to ladder the building," the aide croaked again. "We have fire showing out of all windows, first, second, and third floors. Attempting to conduct primary search. All units heavily involved."

In the short distance we had to travel to Tompkins the fire escalated to a second alarm with an estimated three dead and several injured. We parked astern of six engines and four ladders surrounding the holocaust and dived into the trunk for our helmets and turn-out gear.

"We better do a bang-up physical on this one, Eddie, or the chief'll have our ass," I said. Our new chief was an older marshal named Edwin Sheppard, a stickler for routine procedure. Vincent Canty had been promoted to deputy fire commissioner.

Loping along the pavement toward the fire, Duke picked out a white helmet and recognized a battalion chief known to the department, and probably to his mother, as O'Rourke.

"You got three DOAs, chief?" Duke shouted.

"Worse, Eddie. We're up to six." O'Rourke had that utterly exhausted, soiled, anguished look that I had seen only on New York City firemen and in pictures of coal miners waiting at the pit head after an underground cave-in that had trapped their workmates below.

Water was spilling down to the ground floor of the tenement from the hoses playing on the third level when we pushed into the hallway with our flashlights.

"Holy hell, Eddie, look at this." I had my beam running along the baseboard and up the first stairs. The alligation of the wood was deep and close.

Duke went into the room on the left. "Got it here, too, John. Sonofabitch. And here, and here."

The entire ground floor was gutted, a black shambles everywhere we looked. We kept our flashlights on the floor, circling the rooms. The timber was charred in every room. In five separate spots the charring went right through the floorboards and most of the way through the beams underneath.

We carefully climbed the remains of the staircase, following the path of the fire.

"No landlord, I'd say," said Duke.

"Hell, no; this place was fire-bombed, like a war's going on."

On the third floor, at the door to the Jamiesons' apartment, we saw the more intense searing and blistering of the walls and wood where the fire had banked up, waiting like the monster it was for the door to open so that it could roar on. The door was open, and the lieutenant there told us it had been found that way. Firemen were putting the charred body of an elderly black woman into a canvas bag. We joined the crew in the rear bedroom.

"Oh, Jesus Christ." I looked into that room and felt the sobs rise in my throat. Four children were on the bed like black, shiny, rubber dolls, roasted to the bone. Their mother lay across them, the skin of her back split with ribbons of pink.

I turned away, trying to summon professional calm. Are you there, God? I thought angrily. Are you really there, watching these children get burned to a crisp? Innocent kids who've barely started their lives? I was out in the hallway, my mind clearing. Multiple homicides, I thought. But this time those bastards in court won't make it a circus. Whoever did this, I'm going to lock it up so tight Perry Mason couldn't get them off.

"Eddie," I said, "let's get to work on our notes. I'm radioing the photo unit for a complete coverage."

As dawn came to Brooklyn, Duke went off to Brookdale Hospital to interview the Cox family; I canvassed the street. We both came up empty.

A crowd of neighborhood folk straggled in front of the burned tenement. Duke and I separated, wandering among them in our street clothes.

A small-framed boy with copper skin stood alone by a fence across the street. He was shaking his head and mumbling to himself. I was past him when I caught the words: "Ooh, I'se gonna be next." I paused, dusting off my trouser legs. "I knows they gonna get me next."

I turned to him then. "Hey, man, what do you mean?" He looked up, startled. "There's nothing to fear," I said. "Come over to my car for a minute and talk to me."

The boy followed meekly. I signaled Duke, and the three of us got into the Plymouth.

"What's the trouble, man?" I asked. "Tell us what's the problem."

"It ain't nothing, man, nothing."

"Did you know the people in there, is that it?" said Duke.

"Yeah," replied the boy. "That be's my best friend upstairs there."

"Well, if that's your friend, why don't you tell us what happened?" We had an opening now and I pushed on. "Somebody deliberately set that fire and maybe you know why. Were they mad at your best friend? You have girlfriend problems or school problems or street gang problems? What's the story, man? You know anything you tell us will be kept confidential. We won't tell anybody, and you'll be helping your friend even though he's dead, and you'll be helping yourself. If you're gonna be next, you'd better give us the story. We'll take it slow. What's your name?"

"Tonto."

"Okay, Tonto, what the hell happened?"

Tonto recounted how he and Bones had stolen the car, about the heroin, Joey Lark's ultimatum to The Pistols, and the beating he and Bones took in the schoolyard. That was as far as he could go, but he was convinced the fire was no accident.

"It's gotta be the gang," I told Duke. "Too many fires for one man."

"Right. It's not the style of a man like Lark anyway."

Tonto gave us a full list of the gang members and we consulted on them with the homicide detectives on the case. The local precinct had several cops who specialized in street gangs and these men were dispatched to question members of The Pistols. One member, Reuben (Lemon Tree) Murray, lived a few doors away from the gutted tenement and Duke and I picked him up ourselves.

A homicide lieutenant questioned Murray for the best part of an hour in the interrogation room on the precinct's second floor. He was a good cop, much decorated, but it seemed to me that he was working from a disadvantage. He had had little experience with arsonists and he could not know enough about fires to try a few trip questions. When he emerged he was cutting the air with his hand. "Kid knows nothing," he said. "I think we're chasing after the wrong group."

I had been waiting in the squad room with an overwhelming gut feeling that The Pistols were responsible. The detective sounded so adamant I was astonished.

"Let's not be too quick," I said. "Fire-setters are a funny breed. Gotta take them slow and easy. Mind if I try a few technical questions on him?"

"Be my guest," said the lieutenant. His attitude was offhand.

Duke led the way into the room and we plunked down next to the boy, acting like a couple of weary fire marshals who wanted to clear up a few details and get to hell home.

"This won't take long, man," I said. "Eddie, get Lemon Tree here a cup of coffee."

"Okay, John. We could all use a cup."

"Jesus, some night," I said when Duke left the room. I lit a cigarette and offered the boy the open box. He took one and popped it between his lips.

"Hey, look, pal," I grunted, lighting him up. "We know there was a problem. We know about Joey Lark. Tell us the right story and then we can all go home."

"Lissen, man, I wasn't in on it. Dig?"

"Why did they cut you out? I thought you be's a main man in The Pistols."

"They don't cut me out on anything." The kid was offended. "I be's the lookout."

"Lookout?"

"Yeah, man, we had this meeting in the park 'cause Mr. Lark was going to put bullets up our asses on account of Tonto and Bones stealing his stuff."

"We?"

"Mungo and six of us. We took a vote; beat them to death, Tonto and Bones, or burn them. Everybody said burn. That night we go to Bones's house and I be's the lookout. The others threw matches through the windows."

"Look at this, Lemon Tree." I struck a match on my folder, let it draw a good flame, and tossed it to the floor. The flame died on the arc through the air. The match hit the linoleum smoking, but not burning anything.

My eye caught Duke standing outside the windowed door with the coffee. He knew better than to interrupt unless I signaled.

"The thing about matches," I told the kid, "is that they don't stay alight when you throw them. C'mon, Lemon Tree, cut the bullshit."

"Okay, okay. So they used Molotovs."

"Thank you. That will do it for now. I'd like you to repeat the story in front of my partner. I see him out there with our coffee."

I opened the door. "Hey, Eddie, want to come in? He's made a full admission and he'll repeat it."

The homicide men were taken aback. I had been this route before. It was delicious. Two of them came in with Duke and we heard a fuller version of the story, plus the names of the gang members involved. The street gang cops were radioed to bring them in with their parents. All except Mungo were sixteen, or under, and juveniles.

Duke questioned the young ones in separate rooms with the help of detectives. He was polite to a fault, and the cops got impatient. But Duke knew as much about the fire as the arsonists and he got the admissions.

Hector (Mungo) Wainwright was a different proposition. I sat with him and the homicide lieutenant for thirty minutes or more and nothing came out but jive talk.

"Don't jive-ass us any more, Wainwright." I had had a belly full. "We have the story of what happened. Everybody on the goddam street knows the story."

"I ain't no part of it," he said. "I was with my sister-in-law watching television last night."

Suddenly I leaned over and ripped the fur piece off his collar.

"You ain't gonna be no main man where you're going, Mungo."

The youth fingered the naked denim and looked at me with hate. Then he dropped his eyes and fell into a sullen mood, not saying anything. For a fleeting moment I felt sorry for him—a ghetto kid whose only status in life was a cheap fur collar. The memory of the Jamieson children, stacked like firewood on the bed, killed the sympathy, that and the grinning faces of the other gang boys in the squad room. Duke had them collected on two benches angled between the street windows.

I felt a jab of pain under my ribcage; my bones were leaden. It was five o'clock in the afternoon; Duke and I had been going since a noon stake-out the previous day. I was forty-one years old, forty-one going on sixty. Mungo was beyond reach. I pushed out the door and tagged Duke in the corner.

"Goddam, Eddie. What's going on here? These kids are fourteen, fifteen, sixteen, and they're tougher than grown men. That Lemon Tree character we had in first, got a record that includes rape. Mungo Wainwright in there, he's had rape, felonious assault, armed robbery. Look at them now; they act like they've been on a picnic."

"Cool it, John." Eddie had a haggard, gray look to his face. "I'm beat. We gotta get the show on the road. What did you get from Mungo?"

"Zero."

"No matter. Six corroborating stories. He's gone a million on the admissions and evidence."

Duke went to night court with the juveniles; I went with Wainwright to criminal court. The paperwork took forever and it was nine before I got to the hearing room and filed my complaint with the clerk, a young Puerto Rican with a Van Dyke beard.

"We got a long list tonight," he said pleasantly. "It'll be a wait."

"Please put mine on top, pal, I'm begging you. My whole inside is burning up."

"Sonofagun, marshal." He was looking at me closely. "You gotta face white as a fish belly."

He shuffled my papers near the top of his pile, but it was ten before the judge made the arraignment for arson in the first degree and rejected bail.

I hunched behind the wheel all the way home trying to ease the agony under my ribs. The pain was shooting up and down my back and jabbing into my groin. Nausea rose and fell in my throat.

The lights were on at home and I made an attempt to straighten up when I got out of the car and walk normally to the front door. Flo had seen my lights and the sight of her silhouetted at the door collapsed my manly resolves.

"Flo," I gasped. "I think I'm having a heart attack."

She had me to our family physician's home office in ten minutes. I was diagnosed as suffering a gall bladder attack, probably contributed to by the four packs of cigarettes and forty cups of coffee I had consumed in the past thirty-four hours.

Four days of pill-popping, tomato soup, and ginger ale chased the pains away. Another three days and I was back at work. Duke made his return the same day. His physician said he had never seen a man walk into his office in such a state of emotional and physical exhaustion.

The six juveniles came to trial fairly quickly and were duly convicted of arson and homicide. The judge, in his wisdom, decided that their punishment should be banishment from the city for a period of one year.

I started to rise off the court bench to protest this idiocy, but Duke pulled me back.

"What kind of a deal is that," I snapped.

"They make them every day," said Duke. "We did our job. Sit down and shut up."

The boys did leave the city. We made a point to check their new locations with the precinct in Bedford-Stuyvesant and called the local sheriffs down south. Five went to Tennessee and Mississippi, apparently to grandparents, and the sixth to his brother in Augusta, Georgia. What they had was a vacation; six people burned to death and the boys who committed the murders were given a year in the country.

I knew they returned to the city after the year because they gave evidence at Mungo Wainwright's much delayed trial. Not one detail had gone out of my head and I gave testimony that left the defense without any place to go. Brooklyn District Attorney Eugene Gold caught part of it and wrote me a letter of congratulation. Mungo was sentenced to ten to twenty-five years.

Long before the trial Deputy Commissioner Canty nominated me for the Martin Scott Award, named after a former fire commissioner and recently given each year in our division to the marshal performing the best arson-homicide investigation. I asked Canty to switch the nomination to Eddie Duke, who worked harder than I did on the Tompkins Avenue case, but it was my neck that finally craned over for the placement of the ribbon and gold medallion by Mayor Abraham Beame at a ceremony in Pace College.

A check for $250 went with the medal. I cashed it and dropped off $125 at Duke's house on the way home. He refused to take it, so I left the money lying on his kitchen table and went off to Florida for a vacation with Flo and the kids.

Lying out in the Florida sun, I decided, as many others had before me, that tangible rewards usually come too late in life. By the time you get them you are fully aware of your capabilities and limitations. Endorsement is pleasant but no longer encouraging. I would have much preferred that Duke receive the honor, but then he did not need it either.

Chapter 15

The Case of the Flaming Arrow

IN September, 1973, Jim Susman retired and I was recommended for promotion to supervisor of our nine-man squad. Flo was lukewarm about the idea because I enjoyed working the streets with Duke. This was the factor that made me hesitate too.

One hot Sunday Duke and I discussed the prospect while sipping wine and floating on rubber rafts in my small backyard pool.

"You gotta move up, John, it's your nature," said Duke. "You're a guy who likes his day in the sun. I've seen you on the witness stand. You like to be the main man."

"The way things are is perfect for me," I countered. "I got the best job in the city and the best guy to work with."

Duke splashed his hands idly in the water. "I should have told you before. I'm retiring next year."

I looked at him gratefully. He was lying through his teeth; an active man like Duke was not going to accept half retirement. It was his way of telling me not to be a horse's ass, to grab for the brass ring. By the end of the week I was Supervising Fire Marshal John Barracato and standing dinner for my Group D at Angelo's in Brooklyn.

Reinforced with anisette-spiked coffee, I jumped in Car 55 with Duke and his new partner, Harry Sampson, for the night patrol. We would write a flaming chapter in the history of arson detection this tour, I vowed.

Somewhere along Ocean Parkway we were notified by radio to telephone Brooklyn dispatch. No explanation needed: whatever the problem, it was too delicate to put on the open air waves.

Sampson made the call and came back laughing. "Sounds like an Indian uprising. Party down in Sheepshead Bay says a flaming arrow was shot through the back door."

"Watch it, Harry," said Duke, trying not to crack a smile. "We got the supervisor aboard. He don't fool with anything but major cases. Let's roll in with lights and sirens."

"Right," said Sampson.

"It could be serious," I reminded them briskly.

The address given by the dispatcher was a brownstone with two steps down to the front door. We herded down to the door and knocked.

"Come on in," invited a piping female voice as the door swung open.

There, at our knees, was a dwarf lady, maybe forty inches high, middle-aged. Man, oh, man, I thought, I should have gone with the Bronx team or, even better, stayed at headquarters where I belonged.

We showed our badges and the dwarf introduced herself as Maria Razzetti and led us down a passage to the kitchen, where her husband, a good five feet ten inches, stood waiting.

"Look at this," said the woman, waddling past her husband and producing an arrow from the kitchen bench. Duke inspected it carefully—two feet long, wooden, feathers at one end, a wad of blackened, gasoline-soaked cloth tied at the other.

"I was opening the back door to put the garbage out and see this thing whizzing through the air," said the woman. "It hit the door jamb and fell to the ground with the end burning. I stomped it out while my husband called the Fire Department."

"Aha," Duke muttered in my ear so that it was inaudible to the rest, "Barracato and The Case of the Flaming Arrow."

"Are you having trouble with kids in the neighborhood?" I asked, not daring to look at my late, unlamented partner.

"Only when they're playing hockey on roller-skates outside," said the husband. "Maria grabs the broom then and chases them away."

The comical mind's eye picture of this scene turned me off for a moment. I fiddled around blowing my nose into a handkerchief. Sampson came to the rescue, playing the sharp detective.

"The arrow," he said, "must have been fired by a kid at very close range."

"No, no," said the dwarf lady. "I didn't see anybody. I think it was fired by somebody living across the street."

"Madam," said Sampson, "this arrow could not go through the air with accuracy for any distance because it is too light and the feathers are ruffled."

Duke and I gaped at him. The only arrows Sampson had seen were in the movies.

"Let me demonstrate," continued Sampson, seizing the arrow between thumb and forefinger, aiming it through the door to the living room, and letting fly. We watched in dismay as the arrow went straight and fast, shooting through the air to a beige, flowered couch at the far side of the parlor. The rag at the business end splattered soot over the spotless material.

Sampson got a stricken look on his youthful, tanned face. He leaned down to the dwarf and put his arm on her shoulder. "Gosh. It was accurate after all. I'm sorry, little lady . . ."

The laughter, unkind and uncontrollable, started welling under my collarbone and came up the throat. I swallowed furiously, casting my eyes helplessly about the room. Tears were popping out onto my cheeks. I feigned a coughing spell. Duke had bustled to the kitchen door, looking out, pretending to search for suspicious characters.

With tremendous effort I turned my attention to the woman. She was weeping over the arrow, the couch, the damnfool fire marshals sent to investigate.

"It was very likely a kid's prank," I told her. "But we intend to take the matter seriously. We will question the neighbors and we will question the children in the area. Marshal Sampson will begin with the building across the street, as you suggest."

We retreated from the house and I waited in the car while Sampson went door-knocking in the six-family apartment house facing the Razzetti kitchen. It was the least he could do to appease Mrs. Razzetti. Next day Duke went back to warn off the hockey players and there were no further incidents.

My squad had the arrow mounted on a polished plaque to commemorate my first farcical adventure as a supervisor. It's on my office wall to this day.

In the months following I traveled with all the teams, passing on every trick I had learned since my own early days with Charlie Brewer. Our physical examinations were the most thorough in the division, our questioning of suspects the most patient and fruitful. Our arrest record was unmatched and defense attorneys found it increasingly difficult to break down the mountain of evidence we brought to every case.

Many times I sent marshals back to the scene of a fire to rewrite a vague report on the cause, and how and why the fire spread. On several occasions I interceded in the precinct questioning to show how the honeyed approach worked nearly one hundred percent of the time where harsh tactics failed. "Our job is to get the fire-setters off the street," I said repeatedly. "I don't care if the cops think we're soft-heads just so long as we accomplish this. Turn the other cheek if you have to, act a role, lie a little, but get that hard evidence and the admissions that will stand up in court."

My techniques ran counter to the image of the big-city cop and a few of the marshals—bewitched with the gang-busters image—rejected them as too tame in the beginning. An incident with Brian Rooney, one of my former, forcible partners, swung them my way.

Rooney had picked up a young man named Gregory Lynch on suspicion of setting fires to furniture inside the rectory of a Roman Catholic church in Parkchester, in the Bronx. It seemed a straightforward case. The priest said Lynch, who had worked for him as a general cleaning man, was sacked. Shortly before the fires broke out, neighbors had seen Lynch's white Rambler automobile parked in front of the rectory. The priest was away at the time.

Around 2:30 A.M. Rooney awakened me in the bunk room on Duane Street and asked for help in the interrogation.

"I've had this bastard here for two hours," Rooney told me over the telephone from the precinct in Parkchester. "He's as guilty as hell, but I can't get a peep out of him."

An hour later I walked into a bare office off the squad room and felt my blood rise. Lynch was sitting on a plank seat, naked as a jaybird. He was a lean, sharp-featured man and at this moment he was hunched over, covering his genitals with crossed arms.

"What are you doing, Brian?" I asked irritably.

"I did a strip search and went ahead with the questioning," explained Rooney. "I told him that when he talks he gets his pants back."

Lynch's clothes were stacked in a chair in the corner. I threw them over and told him to get dressed.

Out in the squad room I jabbed a finger into Rooney's thick chest. "Listen, you big Irish lug, don't you know yet that a guy's sense of dignity comes off with his clothes?

You embarrassed him. You made him squirm. You sealed him up. He wouldn't give you the time of day after that."

Rooney shrugged, not really understanding, and filled me in on the case details, including the fact that Lynch had gained entrance to the rectory by putting his fist through a glass panel at the front door. I had noticed a handkerchief wrapped around Lynch's right hand. The cuts, said Rooney, were minor.

A curious detective lieutenant, who had been hovering around, consented to the use of his private office. I took Lynch in there and closed the door, the two of us alone.

"What's the story, Greg?"

"My story is that I was home sleeping with my wife," he said.

"How long you been married?"

"Nine months."

"She's still a bride," I said. "Surely you don't want her to go through a tough time. And tough is what it'll be if you don't trust me to help. We want to be able to tell the district attorney that you were cooperative, that you genuinely want help. You gotta problem, Greg, setting fires like that, and you need help."

"Look, marshal, I realize you're doing your job, but I was out drinking with my brother-in-law early tonight. You can check with him. We both got pissed and went home."

"Greg," I said, constructing my own quiet lie. "You interrupted me. I hadn't finished. You're making all this unnecessarily difficult. I did want that voluntary admission for your sake. See, we found blood on the glass of the rectory door and we're having it analyzed at the laboratory. Then we'll have your blood analyzed. If they match, and they will, it will be conclusive evidence. Whether you speak out now or not, there's no way we're not gonna come out on top."

If Lynch had given a second thought to the blood test, he would realize its absurdity and see it as a desperate last turn of the cards. I gambled that at four o'clock in the morning his reasoning power would be in recess.

"All right," he said wearily. "We had been drinking, but my brother-in-law passed out on the way home. I got thinking about the priest firing me and it got me riled up. I drove over to the rectory to bawl him out. The place was in darkness. I busted in and mussed the place up a bit. There was a stack of newspapers in the closet. I had put them there myself. I tore them up and stuffed them under the chairs and sofa and put a match to them. After that I ran outside and got in the car. I hadn't gone a block before I started to worry that the whole rectory might go up. Nobody told you who pulled the fire box, did they? Well, you're looking at him."

I leaned back in the lieutenant's chair. "Feel better?"

"Yeah, I guess I do," said Lynch.

"Okay, the next step is to repeat the story to Marshal Rooney. He's the one who's gonna talk to the D.A."

"It's a deal," he said, "providing you stay. That other guy is a heavy."

Rooney got his admission and afterwards asked me how I broke Lynch down.

"That's the thing," I said. "I didn't break him down. He could see that I was sympathetic and he opened up little by little. Then I put on the con and he came all the way. He was relieved that he could. You gotta know what's in a man before you can figure how to get it out. Fire-setters have a profound shame. You know that in advance and the only way to get the cause of it, the act of arson, into the open is to be gentle, understanding, to build their trust. Eddie Duke is a master at it. I've seen him crying real tears with a suspect. And, between you and me, I took the job as supervisor because that same sonofabitch dug into me and played on my not too deeply buried vanity."

"The whole thing sounds a bit weak to me," muttered Rooney.

"I don't give a fuck, Brian, about what others think. What counts is the result. When you called me you had nothing but circumstantial; now you got an admitted perpetrator."

Rooney gave me a broad Irish smile. "Jesus, I was just

remembering the Flugelman case we were on three years back. When you walked into the 71 with Alvin Flugelman, the cops looked at you like you had three heads. They said nobody took Alvin before without ten armed men. How did you reach him?"

"There's no connection between that and this." I didn't want to tell him about Flugelman.

"Listen, John, that's a cop-out. You're supposed to be the teacher. We gotta know the wrinkles, bad and good."

Alvin Flugelman was six feet, eight inches tall and weighed 250 pounds. He was very dim and very violent. He was feared in Flatbush.

Rooney and I had investigated a fire that gutted the fourth floor of an apartment house, and neighbors implicated Flugelman because they had heard him bellowing a threat to burn down his girlfriend's house because she refused to date him any more. The girl's name was Harriet, a big, homely person with a faint mustache. Flugelman was crazy for her.

The police told us Flugelman lived with his mother just off Eastern Parkway, and warned against trying to take him alone. They offered to do the job the next day with a full squad.

"You go in your style, John, and he'll crack your head like an egg," Lew Malloy, a homicide man, told me.

"It's my collar, Lew," I said. "And I'll carry it through."

I drove directly to the Flugelman house from Staten Island next morning, composing my act.

Mrs. Flugelman opened the door to my knock and pointed to her son, curled up, sound asleep, on the living room couch. She shook him awake and he got slowly to his feet. He was fully dressed and, sure enough, six feet, eight inches.

"Alvin, I'm from the Fire Department . . ." I began.

He glowered at me from under his overhung brow and motioned in the direction of the kitchen. His mother had gone to make coffee. Obviously he did not want to talk within her hearing.

Out in the car I launched my deceitful play. "Alvin, Harriet is really pissed off at you about the fire. It wasn't a bad fire, of course." (Eight people had been evacuated but I was not about to drop that on him). "But Harriet, she's sensitive, and she's mighty upset."

"Yeah," he said dumbly. "I shouldna burned her out."

"No, you should not. The important thing, though, is that I think I can patch things up between you and Harriet. She's down at the precinct now and she's ready to talk things over. That's why I'm here. I'm a fire department marshal and my job, pure and simple, is to settle the quarrel between you two so that there's no more fires."

"Sounds okay," Flugelman grunted. "I'll give it a shot."

By the time we reached the parkway Flugelman was frowning. "You leveling with me? I mean, why is she at the precinct? I don't wanna go to no precinct. You call and tell her we'll meet at her sister's place."

"Alvin," I said with sudden inspiration. "I'm hungry as a bull. How's about breakfast?"

"Yeah, man." His lips came back and he brayed like a happy horse. We stopped at a diner and I pretended to telephone Harriet while he attacked a double order of fried eggs and ham.

"Looks good, Alvin," I said, returning to the table. "She's already at the precinct waiting for you."

His doubts about the strange rendezvous place had been driven out by the joy of eating. He swallowed down a second cup of scalding coffee and made a fluttering motion with his hand in front of his breast. "Ah-ha," he boomed. "That Harriet is something else. Voo-va-da-voom!"

We marched into the precinct and I led him straight upstairs and along the corridor and into the holding pen. I swung the iron-barred gate shut and turned the key.

"A formality, Alvin," I said. "Let's see what Harriet has to say."

The cops in the squad room, who had seen us pass by, gaped at me when I strolled in. "Came quiet as a lamb," I said grandly.

After stalling around for five minutes I went back to Flugelman. "Tough luck, Alvin, Harriet changed her mind and says she wants you arrested."

"Can't you talk her around?"

"Hey, I already talked until I was blue in the face."

He was not pleased. "Bring the bitch here."

"I can't," I said, backing off a yard from the bars. "She walked out of here so angry I don't think she'll ever come back."

Rooney was aghast when I finished the story.

"That's plain unethical, John."

"I always thought so too," I said. "Then I tell myself, well, he set the fire, I didn't. What did he want from me? I'm no Boy Scout. I go in heavy and he's gonna crush me. No way a .38 would stop him. If a .38 slug can bounce off a car windshield, as they say it can, it would sure as hell bounce off this guy's chest."

If Rooney had an impression of me as a smooth confidence man, passing off trickery as compassion, one young marshal in our group named Pete Garcia had me pegged as an out-and-out cynic. Garcia's problem was that he believed everything he was told, he would assume innocence until he found—or had thrust upon him—a whole sequence of facts that left not the slightest doubt that a person proclaiming his innocence was indeed guilty. I kept telling him to work in reverse, to presume guilt until there were a damn good reason to absolve anyone directly involved of blame.

Our group caught a case in Bushwick that had tremendous instant publicity because a young Hispanic, Raymond Pasquina, made a herioc dash through falling bricks and timbers to rescue a man who had jumped from a burning building and impaled himself on a steel picket fence. The stake had penetrated a good eight inches into the man's abdomen, but excellent fast work by a team of surgeons had saved his life. The wife of the impaled man made a more successful leap from the second floor of the ramshackle two-family house, though she suffered severe burns. The couple's eighteen-month-old son had perished in the blaze.

Bill Farley, a practiced and diligent investigator, conducted the physical of the premises with Pete Garcia. They agreed that flammable liquid had been used in two places to accelerate the fire in the front hall. In the first floor apartment, where the hero had lived, Farley was convinced that two other fires had been set. Garcia was equally certain that the apartment had been reduced to a burned-out hulk by the fire spreading from the hall.

The marshals spoke at length to Pasquina. His story was that he had first become aware of the fire when he awakened to see flames coming through the wall. He dived out the window into an alleyway and ran to the corner to pull the fire alarm box. The fire was raging when he returned and the debris was falling all around the building. He spotted Mr. Tobia, his upstairs neighbor, at the window and then saw him tumbling through the air. Horrified, he watched Tobia fall astride the picket fence. With no thought to his own safety, and this part was confirmed by neighbors, he rushed forward to pull Tobia free.

"Was your door locked at the outbreak of the fire?" Farley had asked.

"Yes, it was," replied Pasquina in Spanish. Garcia was interpreting.

"We found what looked like two separate fires in your place," continued Farley. "We also found empty cans labeled brake fluid."

"What are you saying?" Pasquina demanded. "That I had something to do with the fire? Look, sir, I am a religious man, and I swear to you on the Madonna that what I have just told you is the truth. I do not commit crimes. Never in my life have I been in trouble with the law. The brake fluid, if you must know it, I use in my work. I am an auto mechanic. The shed at the side of the house is where I repair cars. See for yourself."

Garcia had taken the man's side then, and pointed out

to Farley the obvious fact that arsonists were not in the habit of risking their own lives. Furthermore, the deeper scorching at two places in Pasquina's living room could be explained by the spread, and temporary banking of an extremely fierce fire.

That was the state of affairs when Farley got me on the telephone and asked me to come down and make my own independent inspection.

"Will do, Bill," I said. "Meanwhile, run a check on the guy's record. I don't trust people who go the religion route every time they open their mouths."

"You been too long on the street, John," Farley growled.

The two fires in the hall, I discovered, were readable by the rawest recruit. Inside Pasquina's apartment were positive signs that a separate fire had started on the floor at the back of the couch. The floor was abnormally charred and the underside of the couch was almost completely burned out—more so than the cushions on top.

"Look closely at this, Pete," I said. "If this was a regular fire extension, the burning would have been from the outside and top of the couch, inward and downward. This is a reverse situation."

Farley's second suspected apartment fire was against the wall in the right-hand corner of the room. The wall there had been incinerated. There were fissures through to the hall. Garcia was adamant that it was the path of the fire from the hall. A few yards to the right of the damage, however, some cotton-composition drapes were barely scorched, and I asked Garcia how he could explain this phenomenon with his theory of an onward rushing fire. Such curtains would ignite at 400 degrees, the wall ignited at 675 degrees. It had to be a separate flash fire, burning into the wall and traveling upwards. The blast of heat from the central fire would have destroyed the drapes.

To prove the observations Farley went to his car and fetched a hydrocarbon reader, a portable device that inhales residual vapors, runs them over heated coils, and gives a combustion force reading. The more volatile the fumes, the more rapid the oxidation, the higher the reading on the dial. No fumes, no reading. We got a measurement of 500 behind the couch, equivalent to gasoline or brake fuid.

When the needle hit 500 I felt a tug of admiration for Farley. He was into his fifties, the manner of a bank auditor, but he was the sort of professional I ranked as the ultimate. His very first, apparently innocent question to Pasquina, before the suggestion of suspicion, had won an admission that the door to the apartment was locked. This meant Pasquina had exclusive opportunity. We could go to court on that alone.

For Garcia's education we went into the total "cynical" game. The young marshal had thought Pasquina truthful in all things. Our check with the Bureau of Criminal Investigations showed he had two warrants on him for burglary and possession of stolen property. Another lie was established when we canvassed the block. Pasquina had not run to the corner, witnesses said, he had hopped into his car and driven away. Fear of killing someone in the fire had overtaken him, for he did pull the box and he did come back.

We did not get an admission. Motive was established, however, to the satisfaction of the grand jury, because the landlord had days earlier dispossessed Pasquina's common-law wife, the legal tenant, for nonpayment of rent.

The one factor that stuck in our minds about this case, which came down the line finally as strictly routine, was the desertion of the child on the second floor by his mother and father.

We had found on innumerable occasions, and it is worth recording, that people who wake up suddenly to find smoke and flames and heat around them do not think of anyone but themselves. Panic consumes them; self-survival is the dominant impulse. A gallant parent who does dash into the smoke to save a child is quickly overcome by carbon monoxide and therefore ineffectual anyway.

Every marshal puts his family through a fire drill. In my own house we sleep on the second floor. My son John has a little roof spur outside his bedroom window and he is under firm instructions to yell his head off to make sure everyone is awake and then to escape. Under no circumstances is he to try to aid the rest of us. Flo's orders are to jump out the window to the grass below, even at the risk of a broken leg. My job, and I have programmed myself, is to check on Dena and make sure she has opened or busted her window and climbed out on to the roof spur on her side of the house. The last stunt anyone in the Barracato family would pull is to try to run down the stairs, the natural chimney for the fire.

If I ever get to Congress I'll write a bill making it mandatory for every new house in the United States to have a sprinkler system or a heat-sensitive alarm installed at the foot of the stairs—preferably both. From there I shall move on to mass education in the simple precaution of unplugging at night instant-on television sets, toasters, and any other appliances that keep a live current going.

That's a dream, of course, however worthy. I loathe politics.

Chapter 16

Shoot-Out in the South Bronx

THE wave of arson that had been gathering in the late
'sixties and early 'seventies became an epidemic in 1975.
My group was expanded into a sixteen-man major case
squad, on call night and day. As commander, I worked in
liaison with Ralph Graniela, my old street-wise pal,
whose group was posted full time in the South Bronx.
Graniela's assignment was impossible; he was virtually
fighting a war against thousands of arsonists with six men.

The South Bronx was worse than Brownsville had been
at its peak of fury. An average night would bring thirty
fires. The pattern was familiar. The fires were set by
gangs out to cannibalize buildings, by landlords grabbing
for the insurance on their terminally sick tenements, by
the tenants themselves. In effect, the city encouraged the
arson. Anyone who was burned out received compensa-
tion of between one and three thousand dollars for lost
furniture and clothing and moving expenses, plus a priority position on the waiting lists for new public housing.

In a typical case, Graniela and his first lieutenant, Steve
Terco, picked up a kid of fifteen who said he had set fire
to at least forty buildings for payments of about ten dollars

a time from strippers who patrolled the area in trucks. The strippers provided the cans of gasoline but neglected to warn of the dangers. This kid was still surviving; others like him had perished in their own fires. Terco was smart enough to divert the boy from the slipshod juvenile justice system and get him shipped upstate to a home for boys—away from the war zone.

Graniela and Terco worked with the same style that Duke and I had had together. They were totally dependent on each other, dedicated, compassionate. They ventured into the most crime-ridden areas of the city with their tiny .38 guns and thumping hearts, knowing that day by day, arrest by arrest, they were looming as easier targets for the South Bronx hoodlums. Graniela's instinct was the sharpest in the division. He lived by it. But I worried that the charmed lives we marshals had led on the streets of New York would not last forever.

"Ralphie," I told him, "it's getting too hairy. We're sitting ducks. You call me, you understand, at any hour for help. Three in the morning—I don't care when it is. You need us, we're available."

As the spring and early summer of 1975 moved by, Graniela and I continued to hang in by the skin of our teeth. Flames roared in the South Bronx, Bushwick was blazing, and the epidemic was spreading to the seedy lower east side of Manhattan. Newspapers were running major features on the arson menace now. Ministers and priests were thundering on about it from the pulpit. But the city was also going broke; fewer firemen were fighting more fires, the thin division of marshals had scant prospect for acquiring more manpower.

Members of my group, off the regular duty roster, put in sixty hours a week on the serious cases in the five boroughs. One week they scarcely slept at all. A man with a hard, young voice telephoned the Fire Department and my local precinct threatening to burn me out of my own home. Whether it related to a current case or was a man imprisoned on my evidence long before I never knew.

Marshals camped for a week in my living room but the man did not show.

Early in July we toiled around the clock on a series of fires in a skyscraper on Third Avenue. Three lower floors were gutted in one night blaze, which was caught in time by the fire crews. It could have been a "towering inferno," similar to the conflagration depicted in a popular movie; the potential is there in many skyscrapers. Years earlier a fourteenth-floor fire had panicked night workers high up in an office building on Sixth Avenue. They ran to the elevator instead of the stairs. An electrical break swung open the elevator doors at fourteen and flames incinerated four of the ten people inside.

By a long process of elimination, starting with two hundred people, we got to a single maintenance man for the Third Avenue arson. He was remembered at the scene of all fires because he wore a baseball cap. He later admitted setting the fires so as to create a diversion to escape toilet cleaning duties.

Coincident with this investigation by the major case squad, Graniela's team was searching the South Bronx for a Puerto Rican arsonist who had stolen Steve Terco's gun. Graniela and Terco had been staking out an abandoned tenement due to be fired by strippers. Two men ran out of the building, a flash of fire behind them. They split on the sidewalk. Graniela took off after one, Terco went after the other. Terco, a strong, athletic marshal, quickly overtook his man. He spread-eagled him against a wall and began frisking him.

"You got me, marshal," the man grunted. "The blade is down in the boot."

Terco dived for the boot, realized it could be a trick, and straightened up again. Too late. The man had whipped a pistol from his belt and pointed the barrel at Terco's temple. He grabbed the gun from the marshal's hand and bolted into the darkness.

For days afterwards the marshals rousted every known junkie, thief, and gambler in the area, and they made it clear that they would dog their footsteps until the gun was returned. A street cop like Terco would rather lose his badge than his gun. It's a profound loss of face, an invitation for ridicule. Usually we do not frisk a man alone; always we search from the shoulders down. Terco had been forced to break the first rule because his partner had vanished in the opposite direction and the buildings around were crawling with skels who counted Terco as the enemy. His second mistake could be explained by fatigue; he was drained by the constant danger and the endless pursuits through the dark streets and cellars of the South Bronx.

Graniela got the call later in the week; Terco's gun was in a garbage pail outside a Jesus freak storefront on Fox Street. The gun was found, but it was not enough. Graniela and Terco were effective because they had established an indomitability on the street. They had to have the young Puerto Rican who made the snatch. Again they got their anonymous tip. Their man was Luis Carbine, who lived with a girlfriend in her second-floor apartment on 179th Street. Informants said Carbine, a man of twenty-five with a long record of violence, would probably be at the address on Saturday morning.

While Graniela was mobilizing his team for the raid on the apartment, my squad was wrapping up the case against the cleaning man in the baseball cap. I worked with them all through the Friday and stayed on for a full night shift to replace a supervisor who was out sick. It was after ten o'clock on the Saturday morning before I got home. I drank two cups of coffee on the back patio and then Flo ordered me up to bed. For an hour I lay there, tossing restlessly in the heat, before I dozed off.

Just then Graniela was moving on Carbine. Two marshals guarded the front door and fire escape of the building on 179th, two men took the roof, another the front hall. Graniela and Terco moved up the stairs and knocked on the door of Carbine's woman's apartment. She answered with two little children tugging at her dress. Her eyes flicked over the marshal's badges. Carbine, she said, had not been around for a month.

She raised no objection when the marshals went inside and looked through the apartment. In the rear bedroom Graniela came to a closed closet door. He called Terco to his side, then raised his .38 and spun and pulled the door knob with his left hand. Carbine stood inside, a compact figure clad only in his shorts. A Smith & Wesson revolver was clutched in his hand pointing downward.

"Drop the gun, Luis," said Graniela.

Carbine looked at him dully with bloodshot eyes. "Kill me, officer."

"Drop the gun and come out of there," Graniela demanded.

The marshal heard the click as Carbine cocked his piece. It still aimed at the floor. Beads of sweat showed on Carbine's chest, dampening the black mat of hair. It was noon; the city was a July oven.

"Drop the gun," Graniela repeated. He looked at the mat of hair, the young olive flesh underneath, coursing with life. The bullet would hit there, drilling through the man's chest to his heart. Graniela had never shot at a man. Now, at pointblank range, his time had come and he was unprepared. Nothing in the textbooks told him what to do in this situation, gun to gun, his up and aimed, the other's down but cocked—and the man refusing to move.

Graniela reached into the closet with his left hand, intending to break the stand-off by pulling Carbine into the bedroom. At that instant, Carbine's gun came up, shooting. The bullet hit Graniela's wrist, causing his own gun to fire. A red tracer mark showed on the soft flesh of Carbine's left inner forearm. Terco darted forward, a .45 jumping and firing in his hand, the explosion almost simultaneous with Carbine's second shot. Again Carbine suffered only a graze. But his slug pierced Graniela's arm, tearing into his chest, through the lungs, and slamming into his spine.

Graniela looped through the air with a strange sense of

slow motion. Even as his body jerked backward he thought of his good luck in taking out a new $50,000 life insurance policy. He hit the floor, still conscious, aware of everything going on around him, and knew that he was not going to die. His legs went numb, then felt as if they were floating free of his body. He watched Carbine squeezed off a third shot and saw Steve Terco clutch at his throat. Blood gushed through his fingers. Terco put his head against the wall and rolled around the room in agony. The slug had shattered the back of his mouth. Carbine ignored them both, creeping forward to collect the marshals' guns.

Footsteps thumped outside in the hallway of the apartment, It was Arthur Hepburn, who had been waiting on the ground floor. When he got to the bedroom door, Carbine had his pistol at Graniela's head.

"Leave the apartment," he snarled, "or I'll blow both their heads off."

Graniela heard Hepburn's footsteps retreating.

Carbine ran through the apartment to the living room and looked out the window. The fire escape platform was just outside. The gunman saw the marshals on the sidewalk below. He busted out the apartment door and Graniela listened to him pounding up the stairs to the roof in his bare feet. The woman and children had disappeared.

The two marshals up above had heard the shooting, placed a shovel against the scuttle door, and ran to the side of the building to watch the fire escape. The shovel clattered to the asphalt, warning the marshals, and Carbine came through the door. He was still in his shorts, a gun in each hand.

"Halt! Police! Drop your guns!" It was Ray Kirby shouting.

Carbine fired both pistols. The marshals, taking cover behind chimneys, shot back. Kirby cursed his .38 and its two-inch barrel. Several shots caught Carbine in the exchange of fire and he fell sprawling. Kirby was up and charging across the roof when Carbine got to his knees, reaching for the guns that had fallen from his hands. Kirby fired as he ran. Click! Click! The gun was empty. Kirby took a sight on Carbine's head, measured the distance, and let go with a long, lunging drop-kick. His toe punched into the gunman's chin and snapped him over on to his back. The man's arms and chest and legs were spouting blood as if he had been cut with small, sharp stones.

Hepburn, meanwhile, had called a ten-thirteen. The street jammed with patrol cars and the building swarmed with cops . . .

At twelve-thirty the phone rang at my bedside. Through the fog of sleep I recognized the voice of the dispatcher from division headquarters.

"Supervisor Barracato?"

"Yeah."

"Two of our men have been shot in the Bronx. The chief wants all supervisors in immediately."

"Who was shot?"

"Ralph Graniela and Steve Terco."

"How bad?"

"We don't know."

I got out of bed and slowly went to the closet. The calamity had not registered. Flo, who had been napping beside me, asked what was wrong.

"Graniela and Terco were shot in the Bronx," I said matter-of-factly. The words seemed to come from some other person. They gave me the shakes. I put my shorts on backwards. The buttons on my shirt were too big for the holes. I left half of them open.

"Call Eddie Duke and all the others in my squad," I told Flo. "I want them into Church Street right away."

I drove to Duke's place in Brooklyn and he pushed me over to the passenger seat. "It's on the radio," he said. "Ralph and Steve are hurt pretty bad."

My mind was reeling. I imagined Graniela and Terco together saying, "Eddie and John were shot; they're dying." We were interchangeable. I sobbed then and beat my fists on the dashboard for the wounded marshals, for myself, for Duke, for everyone else in our division who worked the midnight, murderous streets, going into buildings with their hearts in their mouths, bluffing it out, taking all that second-rate crap from cops, the city not giving a fuck if you lived or died, and the lousy politicians running around relocating tenants at three thousand bucks a time and telling fire marshals they were lucky to be employed. Every last dollar in the world could not buy a Ralph Graniela if he did not exist—a warm, sweet guy who bled for the lost and lonely and frightened in the goddam city of New York.

At Church Street I dispatched my men to pick up the families of Graniela and Terco. Duke and I drove on to Fordham Hospital. Terco was in surgery, his throat so smashed that doctors were fearful he would never talk again. Graniela was in intensive care, the tubes from half a dozen machines keeping him alive. Surgeons had not assessed the full extent of the injuries, but the x-rays showed spinal damage and they were certain that he would be paralyzed for life from the lower back on down. The prognosis was kept from his wife, Mary—the Granielas had four young children—but when I came upon Ralph's father in a small waiting room I knew that he had heard. He was crying like a baby.

"You and Ralphie were like brothers," he said when I put my arm around him. "You know it will kill him if he can't walk."

I had no answers and I spent the next days in a state of fury and frustration. Once I went to the hospital room where Luis Carbine lay, healthier than either Terco or Graniela.

"Watch out," I whispered to him. "One night I'm gonna come in here and beat the shit out of you."

Graniela was transferred to the new wing at Bellevue, in Manhattan, where they had better facilities to cope with his devastating injuries. Marshals took turns sitting with him through the night. I took the fourth night. As the minutes ticked by into hours there was no other sound but the dry rasp of his breathing, nothing I could watch, or would watch, but the awful resuscitator bag going in and out and the plastic tubes coming out of it and running up his nose. His punctured lungs did not yet have a life of their own.

Bubbles of perspiration popped out on his face and the pale flesh between his eyes clenched like a fist. Christ, Ralphie, I thought, what's going on here with the king of the streets? His eyes came open and he looked at me in the dim light of the hospital night lamp with bright terror darting for a moment and then clouding back into a blunt fear.

He motioned for the pencil and scratch pad on the bedside table. I held the pad in my hand while he wrote in large letters: "I was dreaming I was shot again."

"Hey, no way, pal. We're all here. The whole goddamed department is here."

His eyes blinked in response but the fear stayed. "John, hold my hand," he wrote.

Instead of replying I reached over with my free hand for the little blue rubber bag of ice on the table and started gently to dab at his brow.

"I won't tell anybody," he wrote again.

He let the pencil fall and pushed his right hand along the sheet by his side. I returned the ice bag and pad to the table and slipped my hand in his, gripping hard, not wanting to let go. We stayed like that the rest of the night, supervising fire marshals Ralph Graniela and John Barracato, top-ranked detectives in the New York Fire Department, scourge of arsonists, protectors of the city, adventurers in the streets.

Graniela was holding on to life as he plunged in and out of sleep and pain and terror, and I was holding on for times past, making a mute, confused plea that they not be allowed to end here on this ghostly plateau of the Bellevue intensive care ward, shattered by the bullet in Graniela's spine. Our special pride had been to put ourselves above the mean peril of the streets. In our own way, we loved and honored them, and so we endured. We had endured. I wondered through that long night whether we had been living out some boyish fantasy, and that now the game was up we had better return to the arson wars with stouter equipment and calloused hearts. Our marvelous bravado surely was ashes to Graniela now.

By morning light I withdrew my hand and fought against a gloom that had seeped right into my bones. I was glad to hear Bill Farley's footsteps on the tiled floor leading to Graniela's alcove and the fresh, cheerful voice saying, "How's the boy?"

Graniela slanted his eyes over to the table and lifted his hand in writing motion for Farley to fetch the pad and pencil.

"Sonofagun, he didn't?" Farley said in mock horror when Ralph had finished.

I leaned over and read the words: "John held my hand all night."

"You bastard," I said. "You promised you wouldn't tell."

The faintest of grins touched his face.

I took my leave then and went out into the corridor and down to the elevators, nodding to the men and women who whisked by in white and green gowns. Down by the water fountain someone had left a silver cart unattended and I took an apple from a fruit bowl that was peeping from beneath the covers.

Jesus, that Graniela, I thought. Always the punchline. Did they know out there in the streets what they had lost?

During the hot summer weeks after the shooting many of the marshals agitated for a change in our low-key methods, a return to the bare-knuckle Charlie Brewer days. And they wanted stronger protection against the thousand and one quick-triggered skels who were free to roam our streets because courts were crowded and jails were packed.

But I was determined that one crazy man in a closet would not destroy the humanity and professionalism that Graniela and I had done so much to establish in our division. I put my views to the major case squad. It took a while but they prevailed.

"Nothing will be different on the streets," I told them. "We're still gonna be targets and we're still gonna be entering skel buildings and knocking on doors. We'll be thinking of what happened to Ralph and Steve, and that's natural. But the biggest mistake any of us can make is to panic. We are not gonna start shooting and asking questions afterwards. When the situation does arise, as it did with Ralphie, don't hesitate. Take the necessary action. But be sure of what you're shooting at. There'll be no shooting people for nothing."

A big, florid-faced marshal named George Burton was the first to speak up. "Listen, John, we're well aware of this turn-the-other-cheek stuff you're always pitching, but we got a different ball game now. I, for one, am gonna start carrying a 357 magnum. I'll keep the .38 for shooting holes in cheese."

"You carry one of those cannons, George, and the next thing you'll want a shotgun in the car," I said. "Pretty soon you'll be relying on firepower instead of plain street savvy. That ain't no way to get the fire-setters and it's got nothing to do with getting evidence for the grand jury."

"Why always the kid gloves, John?" Brian Rooney cut in. "The streets are full of animals and you're almost kowtowing to them."

"Think about it, Brian; eight million people in the city. How many would you count as animals? A lot, sure, but still a small minority. We work for the majority and we do that by getting the horrors off the streets. Guns won't do it. We gotta build cases, it's the only way to go."

We had two cars and the surveillance van on duty in the South Bronx through the summer. We shook down the informers and two-bit hustlers for scraps of information. We sat on rooftops and mooched as deadbeats and skulked in basements and dogged the fire engines. We made sixty arrests of landlords, strippers, and hired torches, and reduced the fires by half. It was accomplished without firing a shot, and talk about toting magnums ceased. Yet the thought was rarely out of my mind that real success would come only if marshals and cops waged the arson war in tandem and with mutual respect.

Arson was up two hundred percent in ten years and we were trying to cope with fifty-three marshals, less than when I joined the division. We had the expertise; the cops had the manpower and the intimate knowledge of local neighborhoods and the felons on the loose within them. The Police Department should be assigning detectives to us for training, in my opinion, and they should be working out of our division. Cops knew about thieves and murderers, but arsonists were a different breed. When I asked homicide detectives if they thought I could run a thorough murder investigation, they looked at me as if I had rocks in my head. It is the same with you guys investigating arson, I told them. I remembered Brownsville and Larry Goldman, and our powerful scores together, but nobody wanted to be a Goldman. The cops wanted no part of us. The Fire Department was a blind spot with them. They worked on arson cases apart from us, even though they were totally unqualified for such basic detection as whether a fire was accidental or set intentionally. And many cops were reluctant to arrest suspected arsonists anyway because the crime was so damned hard to prove in court unless you were an expert.

The foolish rivalry had to be broken and it became my crusade and a compelling reason to accept a promotion to deputy chief fire marshal in September, 1975. The assignment meant moving behind the war lines to an office and a desk, and from there directing the city's main assault on arson.

It was not my style to be inside, away from the streets. On my first morning I felt alone and nostalgic for years past. Many of the episodes narrated in this book went through my mind in bright, quick pictures. I prickled with envy for the marshals out there prowling the city that misty Manhattan morning, but I shook it loose finally. I picked up the phone and set up interviews in the Bronx district attorney's office and the precincts for the purpose of arguing my case to get cops and marshals on the same team. The men at the other end of the line were noncommittal but friendly. My new title carried unexpected clout. It would be a long, arduous campaign, and maybe it would take a little con. The prospect did not dismay me.

Meanwhile, however, the city was still burning.

I sent out for a sandwich and worked out a fresh interlock patrol for the South Bronx, sixteen men in four cars. Against the lead night car I wrote the name Barracato. I sat back, looking through the windows to the towers of Wall Street, and wondered how the Commissioner would take it if his new chief spent a few weeks, just a few short weeks, back on the streets.